Inverse and Ill-Posed Problems Series

Managing Editor

Sergey I. Kabanikhin, Novosibirsk, Russia / Almaty, Kazakhstan

Vladimir V. Vasin
Ivan I. Eremin

Operators and Iterative Processes of Fejér Type

Theory and Applications

Walter de Gruyter · Berlin · New York

Authors

Vladimir Vasilievich Vasin
Institute of Mathematics and Mechanics
Ural Branch of the Russian Academy of Sciences
16, S. Kovalevskaya street
620219, Ekaterinburg GSP-384
Russia
E-mail: vasin@imm.uran.ru

Ivan Ivanovich Eremin
Institute of Mathematics and Mechanics
Ural Branch of the Russian Academy of Sciences
16, S. Kovalevskaya street
620219, Ekaterinburg GSP-384
Russia
E-mail: ermii@imm.uran.ru

Keywords:

Fejér mappings, ill-posed problems, gradient methods, Levenberg–Marquardt method, α-processes, parallelization, minimal error method, nonsmooth optimization, convex inequalities, Newton–Kantorovich method, randomization

Mathematics Subject Classification 2000:
65-02, 65Jxx, 65Kxx, 65R32

∞ Printed on acid-free paper which falls within the guidelines of the ANSI to ensure permanence and durability.

ISBN 978-3-11-021818-3

Bibliographic information published by the Deutsche Nationalbibliothek

The Deutsche Nationalbibliothek lists this publication in the Deutsche Nationalbibliografie; detailed bibliographic data are available in the Internet at http://dnb.d-nb.de.

Printed in Germany
Cover design: Martin Zech, Bremen.
Printing and binding: Hubert & Co. GmbH & Co. KG, Göttingen.

Preface

The book presents some results of the authors' investigations in the area of iterative methods for approximating the fixed points of operators possessing a property of quasi-contractivity (the *Fejér property*). There exists a huge number of references devoted to investigations on convergence of iterative processes for finding the fixed points under various assumptions about classes of operators and spaces. Therefore, it is a very difficult matter to write even a glance review over this literature.

The principal difference of the material presented in this book from the overwhelming majority of works on the topics is in the following:

1) the authors completely concentrate on the investigation of problems in Hilbert spaces and, even mainly, in the finite-dimensional Euclidean space $\mathbb{R}^n$;

2) the main focus is not on obtaining the most generalized theorems on convergence, but rather on constructing concrete mappings; on the basis of these mappings, iterative processes are generated that approximate solutions of a wide set of proper and improper problems of mathematical programming, well- and ill-posed problems with presence of additional *a priori* constraints;

3) for constructing iterative procedures, the authors widely use one very important property of the class of the quasi-contractive mappings; this property consists in their closure with respect to superpositions of some type; this provides an opportunity to implement a natural decomposition of the problem and to decompose the algorithm into some simpler procedures.

The quasi-contractivity of operators is met in various forms in many works (see, for example, (Martinet, 1972; Maruster, 1977; Moreau, 1965)) and are coupled with investigations of theorems on existence and approximations of fixed points. Here, the terminology is not established yet. By this reason, in Section 1 of Chapter I we give definitions of the main classes of nonlinear mappings using for this two terminologies. The origins of this terminology are, from one side, in works from numerical functional analysis and, from the other side, in works from the mathematical programming, in which the term "Fejér mapping" (named such in the honor of the outstanding Hungarian mathematician L. Fejér, 1880–1959) is widely known. The origin of this term, which we prefer in this book, has its own history discussed below.

Fejér had introduced the following definition (Fejér, 1922).

Let M be a closed set of points in $\mathbb{R}^n$. If p and p_1 are points in $\mathbb{R}^n$ such that $\|p-q\| > \|p_1-q\|$ for any $q \in M$, then p_1 is said to be pointwise closer to M than p. If p is such that there is no point p_1 which is pointwise closer to M than p, then p is called the closest point to the set M.

He made the interesting observation that the set of points closest to M is identical to the closed convex hull $\overline{\text{conv}}(M)$ of the set M. From this property it follows that if a point p does not belong to $\overline{\text{conv}}(M)$, then it is possible to find a point p_1 which is closer to M than p.

On the base of this geometrical fact, in the work (Motzkin and Schoenberg, 1954) the notion of a Fejér monotone sequence x_k (with $x_k \neq x_{k+1}$, $\|x_{k+1} - q\| \leqslant \|x_k - q\|$ for any k and $q \in M$) was introduced and the relaxation method was constructed for the approximate solution of systems of linear inequalities. So the Fejér approach had been applied to solve systems of linear inequalities. Later, these techniques were developed for solving problems of convex programming.

This property happened to be rather important, and it appeared in investigations of problems in which the set of their solutions plays the role of the set M. For instance, it can be the solution set of a system of linear equations, the set of optimal solutions for problems of linear programming, of ill-posed problems of mathematical physics, and of many others. These all are problems, in which non-uniqueness of solutions is a typical situation.

The problems mentioned occupy the end positions in the historical chain of development of mathematical simulation problems and of expanding the classes of problems to be solved. We speak about a chain of classes with their peculiarities: the uniqueness of solution and its stability (well-posedness) $\longrightarrow$ nonuniqueness and instability (ill-posedness) $\longrightarrow$ insolvability (in a more general case, improperty $\longrightarrow$ poor formalizability).

For such problems new and innovating approaches are necessary on simulating and solving the problems related, in particular, with the synthesis of techniques and methods elaborated in the classic settings.

The notion of Fejér convergence has played an important role. As it was mentioned above, in the beginning it was used in the works (Motzkin and Schoenberg, 1954; Armon, 1954) for the construction of iterative methods for solving linear inequalities. Somewhat later, investigations of methods of such a type were carried out in (Merzlyakov, 1962; Bregman, 1965; Gurin, Polayk, and Raik, 1967; Eremin, 1965a). In the next works (Eremin, 1965a, 1966b, 1966c, 1968a, and others), more generalized notions and terms connected with the Fejér's name were introduced. Those were: Fejér mappings (one- and multi-valued), Fejér methods, and others.

Large number of theorems and auxiliary facts were proved that gave a wide range of possibilities of constructing iterative methods for solving problems of various types: systems of convex inequalities and problems of convex programming, ill-posed problems of mathematical physics in the presence of additional functional constraints, and so on.

Let us pay attention to another circumstance related to the iterative nature of the Fejér operator, for example T, which is put into correspondence to the problem of finding the solution of a system of linear and convex inequalities. The iterative process $x^{k+1} = T(x^k)$, $k = 0, 1, 2, \ldots$ can be convergent even the mentioned system of inequalities is inconsistent. In this case the limit has an approximative meaning. Such cases are quite important, and their identifications and investigations is an actual topic. Some cases of that type are presented in Sections 2, 3, and 6 of Chapter IV.

To avoid ambiguity, let us make one note on a phrase often met in the text of this book: "... let a set M be given ..." from $\mathbb{R}^n$ or $\mathcal{H}$. This or that identification of this set can be either simple or difficult. In this sense it is possible to speak about *explicit* and *implicit* presentations of this set. For example, let the set $M \subset \mathbb{R}^n$ be a polyhedron that can be given in two ways: $M = \operatorname{conv}\{p_j\}_{j=1}^m$ or $M = \{x : Ax \leqslant b\}$.

The first presentation is explicit and has the following properties: it is *simple* to find $\bar{x} \in M$, but it is *difficult* to check $\bar{x} \in M$.

The second way is implicit and has the following properties: it is *difficult* to find $\bar{x} \in M$, but it is *simple* to check $\bar{x} \in M$.

These two ways of presentation are dual. They can be seen in other prob lems. Since practically everywhere in the text the set M, meaningfully, is an element of some object to be found, it is necessary to give the sense of the implicit presentation to the set M. Note that the above mentioned terms "simple" and "difficult" are used not in the exact sense, but in a pithy indirect meaning. It is easy understood in the example given above in the case of an analytical description of the (bounded) polyhedron from $\mathbb{R}^n$.

Now briefly consider the contents of the book.

In Section 1 of Chapter I the main classes of nonlinear mappings (of Fejér type) are defined, and their general properties are studied. Further, these properties are used for construction and investigation of iterative processes. In particular:

– the very important property of closure of these classes with respect to some transformations (multiplication and convex summation) is established;

– theorems on the strong (Section 2) and weak (Section 3) convergence of the successive approximation method with step-operators of the Fejér type are formulated, and meaningful applications are discussed;

– the Browder fixed point principle (Browder, 1967) is considered for nonexpansive operators, and the strongly convergent method of the correcting multipliers is formulated (Subsection 4.2, Section 4); the idea of this method belongs to Halperin (Halperin, 1967);

– the asymptotic rule for stopping the iterations is defined, and this rule guarantees the stable approximation of solutions of ill-posed problems for approximately given input data (Subsection 4.3, Section 4).

In the last three sections of this chapter, Fejér mappings and processes are investigated both for one-valued and multi-valued operators, the approach of constructing such mappings is described on the basis of dividing couples, and the basic constructions of Fejér mappings are discussed.

In Chapter II, linear and nonlinear operator equations in Hilbert spaces are discussed. For linear equations, properties of an iterative operator in the α-process is analyzed, whose concrete realizations lead to the methods of steepest descent, minimal residuals, and minimal error (Section 4). In the nonlinear case, the following methods are investigated: Newton–Kantorovich (Section 2), Levenberg–Marquardt, and linearized versions of the gradient methods (Sections 5 and 6). The local (in the neighborhood of a solution) conditions on the original operator are formulated, which are sufficient for the step-operator to belong to the class of the pseudo-contractive mappings that, in general case, guarantee weak convergence of iterations. The superposition principle for operators of Fejér type allows one to solve not only the operator equations, but, also, systems of nonlinear equations together with a system of convex inequalities (Section 3).

In Section 7, the considered methods are applied to ill-posed problems having additional *a priori* constraints when the solution iterative operator is constructed as the superposition of some classical method, for example of one mentioned above, and a Fejér mapping, which is responsible for the constraints. Several applied problems are described that are reduced for solving integral (one- and two-dimensional) equations of the first kind with *a priori* information on the solution, and the results of numerical experiments are discussed from the viewpoint of effectiveness of the suggested method.

In its essence, Chapter III illustrates and deepens the contents of Sections 5–7 of Chapter I by examples of objects, which are very important for mathematical programming: systems of linear and convex inequalities both consistent and inconsistent. In Section I the basic constructions of Fejér mappings are considered for systems of linear and convex inequalities. The case of the mirror relaxation is discussed in the Fejér process for a system of linear inequalities (Section 2). In Section 4 the important case of the Fejér process is investigated for a finite system of convex sets with an empty intersection. The convergence of the cyclic process of projecting to the cycle of immobility is proved. Section 6 is devoted to aspects of the topic "Fejér methods and nonsmooth optimization". The aspects of the convergence rate of the Fejér processes (Section 5) and their stability (Section 7) are considered.

Chapter IV has six sections and presents some special topics associated with the Fejér processes, in particular: methods of their parallelization (Section 1), randomization (Section 2), the Fejér processes for consistent and inconsistent systems and improper problems of linear programming of the first, second, and third kind (Sections 3–4). In Sections 5–6 some topics of procedures for the problem of point projection onto a convex closed set from $\mathbb{R}^n$ are considered.

The authors highly appreciate the work of colleagues from the Institute of Mathematics and Mechanics, Ural Branch of Russian Academy of Sciences (RAS), L. D. Popov and B. D. Skarin, who read the book in preparing the Russian publication and provided useful comments. The authors express deep gratitude to G. F. Kornilova who provided a computer preparation of the manuscript and spent many efforts for its improvement.

This book is the translation of a monograph edited in Russian by the Ural Brach of RAS (Ekaterinburg, 2005).

April 2009 V. V. Vasin, I. I. Eremin

Contents

Introduction

The book suggested to an interested reader is devoted to iterative methods for solving some classes of problems (ill-posed ones, linear operator equations, systems of linear and convex inequalities, problems of linear and convex programming) that are generated by operator of the nonexpansive types. In the latter types we consider strongly M-nonexpansive (or M-Fejér), quasi-nonexpensive, pseudo-contractive, and some others.

The minimal and natural requirement on any iterative process is the decreasing of the norm of deviation of the iteration point from the solution to be found or from the set of solutions as the number of iteration increases. This requires the condition of strong nonexpansibility (or Fejér-type property) onto the operator. The need to obtain convergence of iterations to some solution in a sufficiently wide space leads to a stronger requirement onto the step operator that can be expressed by some property of pseudo-contractivity.

The M-Fejér property of a mapping $T : \mathcal{H} \to \mathcal{H}$ (here, $\mathcal{H}$ is a Hilbert space) means that the image $T(x)$ of an element $x \notin M$ lies closer to any point y from M, and thus $M = \text{Fix}(T)$. Firstly, note the constructive character of generating such operators in application to the problem mentioned above; second, the logic and arithmetic structure of such operators is rather simple and has wide internal parallelism that is important for solving problems on contemporary multi-processor computers on the basis of parallelization.

The importance of such operators lies in the fact that the Fejér property (pseudo-contractivity) remains valid under constructing their superpositions and convex combinations. This provides a large range of possibilities for constructing various, rather flexible and economic procedures for solving problems of different nature that are reduced to finding a fixed point of such a mapping. Here, it is necessary to note the universality of these procedures that comes from the fact that insolvability (improperty) of a problem is not an obstacle for their application. In such a case, approximation of some generalized solution holds, and the solution is found by a rather natural way.

Moreover, there is another reasoning that underlines importance of investigations of such operators. Namely, when solving ill-posed problems, it is crucial to take into account all additional information on the solution to be found. The most actual is accounting *a priori* constraints in the case of nonuniqueness of a solution, since its localization gives an opportunity to separate the solution satisfying the properties following from the physical sense of the problem. It is turned out that a wide number of *a priori* constraints arising in applications can be taken into account in a very simple and flexible way by means of special Fejér mappings.

It allows to construct iterative processes on the basis of a modification of classic schemes by means of appropriate mappings of the Fejér type that is responsible for the *a priori* relations described, for instance, by a system of linear or convex inequalities. Examples given in Chapter II demonstrate high efficiency of such an approach to solving important ill-posed problems with simple *a priori* constraints in the form of linear inequalities.

By authors' opinion, the solution of problems on the basis of iterative constructions of Fejér type now has no noticeable propagation in contemporary computational mathematics. The authors hope that systematic description of the Fejér processes implemented in this book will facilitate more wide implantation of this techniques into practice of numerical analysis.

Chapter I
General properties of nonlinear operators of Fejér type

1 The main classes of nonlinear mappings

Let $\mathcal{X}$ be a linear normed space and $T : \mathcal{X} \to \mathcal{X}$ be a mapping (operator) acting in this space. We use the notation $\mathrm{Fix}(T)$ for the set of fixed points of the mapping T, i.e.,

$$\mathrm{Fix}(T) = \{x : x \in \mathcal{D}(T),\ T(x) = x\},$$

where $\mathcal{D}(T)$ is the domain of definition of the operator T in the space $\mathcal{X}$. In some cases for convenience and brevity of writing, we use the denotation M instead of $\mathrm{Fix}(T)$ for the set of fixed points. In general $\mathcal{D}(T)$ is not assumed to coincide with $\mathcal{X}$; nevertheless, if it does not cause ambiguity, we also use the notation $T : \mathcal{X} \to \mathcal{X}$ instead of $T : \mathcal{D}(T) \subseteq \mathcal{X} \to \mathcal{X}$ and "$\forall\, x \in \mathcal{X}$" instead of "$\forall\, x \in \mathcal{D}(T) \subseteq \mathcal{X}$".

1.1 (Quasi-)nonexpansive and pseudo-contractive operators

Definition 1.1. A mapping $T : \mathcal{X} \to \mathcal{X}$ is called *weakly M-Fejér* or *M-quasi-nonexpansive*, if $M = \mathrm{Fix}(T) \neq \varnothing$ and

$$\|T(x) - z\| \le \|x - z\| \quad \forall\, x \in \mathcal{X} \quad \forall\, z \in M. \tag{1.1}$$

Denote the class of such operators by $\mathcal{K}_M$.

Definition 1.2. A mapping $T : \mathcal{X} \to \mathcal{X}$ is called *M-Fejér* or *strictly M-quasi-nonexpansive*, if $M = \mathrm{Fix}(T) \neq \varnothing$ and

$$\|T(x) - z\| < \|x - z\| \quad \forall\, x \in \mathcal{X}, \quad x \notin M, \quad z \in M. \tag{1.2}$$

Denote this class by $\mathcal{F}_M$.

In the case of a Hilbert space we use the denotation $\mathcal{H}$ instead of $\mathcal{X}$.

Relations (1.2) which hold for an operator of metric projection onto a convex closed subset M of a Hilbert space $\mathcal{H}$, are characteristic; i.e., if in a normed strictly convex space the operation of metric projection satisfies the relations (1.2), then this space is a Hilbert space as V. S. Balaganskii had shown.

Definition 1.3. Let $\mathcal{X} = \mathcal{H}$ be a Hilbert space. A mapping $T : \mathcal{H} \to \mathcal{H}$ is called *M-pseudo-contractive* (or strongly M-Fejér), if $M = \mathrm{Fix}(T) \neq \varnothing$ and there exists a constant $\nu > 0$ such that

$$\|T(x) - z\|^2 \leq \|x - z\|^2 - \nu \, \|T(x) - x\|^2 \quad \forall\, x \in \mathcal{X} \quad \forall\, z \in M. \tag{1.3}$$

Denote this class by $\mathcal{P}_M^{\nu}$.

It is worth to note that for $\nu = 1$, inequality (1.3) is equivalent to the following one:

$$\|T(x) - z\|^2 \leq (T(x) - z, x - z), \tag{1.4}$$

where $(\cdot, \cdot)$ denotes the inner (scalar) product in $\mathcal{H}$.

Definition 1.4. A mapping $T : \mathcal{X} \to \mathcal{X}$ is called *nonexpansive* (*nonexpanding*), if

$$\|T(x) - T(v)\| \leq \|x - v\| \quad \forall\, x, v \in \mathcal{X}. \tag{1.5}$$

We denote this class by $\mathcal{K}$.

Definition 1.5. Let $\mathcal{X} = \mathcal{H}$ be a Hilbert space. A mapping $T : \mathcal{H} \to \mathcal{H}$ is called *pseudo-contractive* if there exists a constant $\nu > 0$ such that the following inequality holds:

$$\|T(x) - T(v)\|^2 \leq \|x - v\|^2 - \nu \, \|x - T(x) - (v - T(v))\|^2 \quad \forall\, x, v \in \mathcal{H}. \tag{1.6}$$

We denote this class by $\mathcal{P}^{\nu}$.

It is evident that if $M = \mathrm{Fix}(T) \neq \varnothing$, then for $v = z \in M$ inequality (1.5) transforms into (1.1), and inequality (1.6) transforms into (1.3). Thus, in this case $\mathcal{K}_M \subseteq \mathcal{K}$, $\mathcal{P}_M^{\nu} \subseteq \mathcal{P}^{\nu}$.

Since relation (1.2) implies (1.1) and relation (1.3) implies (1.2), the following inclusions hold:

$$\mathcal{P}_M^{\nu} \subset \mathcal{F}_M \subset \mathcal{K}_M,$$

and, as some examples show (see, for instance, (Vasin and Ageev, 1993)), these inclusions are strict.

Denote the identity operator by I.

Lemma 1.6. *If $T : \mathcal{H} \to \mathcal{H}$, $T \in \mathcal{P}_M^{\nu}$, then $T = \frac{1}{1+\nu} V + \frac{\nu}{1+\nu} I$, where $V \in \mathcal{K}_M$. Conversely, if $V \in \mathcal{K}_M$, then for $\lambda \in (0,1)$ the mapping $T = \lambda V + (1 - \lambda) I$ belongs to the class $\mathcal{P}_M^{\nu}$ for $\nu = (1 - \lambda)/\lambda$.*

Proof. It is reduced to a direct check. For example, let us verify that the following inclusion holds: $\lambda V + (1-\lambda) I \subset \mathcal{P}_M^\nu$ for $\nu = (1-\lambda)/\lambda$. For $x \in \mathcal{H}$ and $z \in M$ we have

$$\begin{aligned}
&\|\lambda V(x) + (1-\lambda)x - z\|^2 - \|x - z\|^2 \\
&\qquad + \frac{1-\lambda}{\lambda} \|x - (\lambda V(x) + (1-\lambda)x)\|^2 \\
&\quad = \|\lambda (V(x) - z) + (1-\lambda)(x - z)\|^2 - \|x - z\|^2 \\
&\qquad + \lambda (1-\lambda) \|(x - z) + (z - V(x))\|^2 \\
&\quad = \lambda (\|V(x) - z\|^2 - \|x - z\|^2) \le 0,
\end{aligned}$$

since $V \in \mathcal{K}_M$. □

The analogous statement holds for the classes $\mathcal{K}$ and $\mathcal{P}^\nu$.

Lemma 1.7. *If $T : \mathcal{H} \to \mathcal{H}$, $T \in \mathcal{P}^\nu$, then $T = \frac{1}{1+\nu} V + \frac{\nu}{1+\nu} I$, where $V \in \mathcal{K}$. Inversely, if $V \in \mathcal{K}$, then for $\lambda \in (0,1)$ the mapping $T = \lambda V + (1-\lambda) I$ belongs to the class $\mathcal{P}^\nu$ for $\nu = (1-\lambda)/\lambda$.*

Thus, for $0 < \lambda < 1$, the operator $T_\lambda = \lambda T + (1-\lambda) I$ satisfies a stronger contractivity condition than the original operator T, and in addition $\operatorname{Fix}(T) = \operatorname{Fix}(T_\lambda)$. We shall widely use these properties in the study of iterative processes.

Theorem 1.8. *Let $T_i : \mathcal{H} \to \mathcal{H}$, $T_i \in \mathcal{P}_{M_i}^{\nu_i}$ and $M = \bigcap_{i=1}^m M_i \neq \varnothing$. Then:*

1) $T = T_m T_{m-1} \dots T_1 \in \mathcal{P}_M^\nu$, *where* $\nu = \min\limits_{1 \le i \le m} \{\nu_i\} / 2^{m-1}$;

2) $T = \sum\limits_{i=1}^m \alpha_i T_i \in \mathcal{P}_M^\nu$, *where* $\alpha_i > 0$, $\sum\limits_{i=1}^m \alpha_i = 1$, $\nu = \min\limits_{1 \le i \le m} \{\nu_i\}$.

Proof. First, consider the operator $T = T_m T_{m-1} \dots T_1$. It is evident that $M \subseteq \operatorname{Fix}(T)$. Let us verify that the inverse inclusion is valid. For this, it is sufficient to consider the case $m = 2$. Let now $x \in \operatorname{Fix}(T)$, where $T = T_2 T_1$. Assume the contrary, namely, that $x \notin M_1 \cap M_2$. Then two cases are possible: 1) $x \notin M_1$ and 2) $x \notin M_2$, $x \in M_1$.

In case 1) for $z \in M$ we have

$$\begin{aligned}
\|x - z\|^2 &= \|T_2 T_1(x) - z\|^2 \le \|T_1(x) - z\|^2 - \nu_2 \|x - T_2 T_1(x)\|^2 \\
&\le \|x - z\|^2 - \nu_1 \|x - T_1(x)\|^2 < \|x - z\|^2,
\end{aligned}$$

and in case 2)

$$\begin{aligned}\|x - z\|^2 &= \|T_2T_1(x) - z\|^2 = \|T_2(x) - z\|^2 \\ &\le \|x - z\|^2 - \nu_2\|T_2(x) - x\|^2 < \|x - z\|^2,\end{aligned}$$

i.e., we obtain a contradiction.

For the operator $T = \alpha_1T_1 + \alpha_2T_2$, the assumption that $x \in \operatorname{Fix}(T)$, $x \notin M_1 \cap M_2$, also leads to the following impossible inequality:

$$\begin{aligned}\|x - z\| &= \|\alpha_1T_1(x) + \alpha_2T_2(x) - z\| \\ &\le \alpha_1\|T_1(x) - z\| + \alpha_2\|T_2(x) - z\| < \|x - z\|,\end{aligned}$$

since $\mathcal{P}_{M_i}^{\nu_i} \subset \mathcal{F}_{M_i}$, and we have either $x \notin M_1$, or $x \notin M_2$. Now discuss the proof of the fact that $T = T_2T_1 \subset \mathcal{P}_M^{\nu}$. Applying the theorem condition that $T_i \in \mathcal{P}_{M_i}^{\nu_i}$ $(i = 1, 2)$, and convexity of the square of the norm, we have for $z \in M = M_1 \cap M_2$, $x \in \mathcal{X}$:

$$\begin{aligned}\|T(x) - z\|^2 &= \|T_2T_1(x) - z\|^2 \\ &\le \|T_1(x) - z\|^2 - \nu_2\|(I - T_2)T_1(x)\|^2 \\ &\le \|x - z\|^2 - \nu_1\|x - T_1(x)\|^2 - \nu_2\|(I - T_2)T_1(x)\|^2 \\ &\le \|x - z\|^2 - (\min\{\nu_1, \nu_2\} / 2)\, \|x - T_2T_1(x)\|^2,\end{aligned}$$

i.e., $T = T_2T_1 \in \mathcal{P}_M^{\nu}$, where $\nu = \min\{\nu_1, \nu_2\} / 2$.

If $T = \alpha_1T_1 + \alpha_2T_2$, where $\alpha_i \in (0, 1)$, $\alpha_1 + \alpha_2 = 1$, then for any $z \in M = M_1 \cap M_2$ and $x \in \mathcal{X}$, the following relations hold:

$$\begin{aligned}\|T_1(x) - z\|^2 &\le \|x - z\|^2 - \nu_1\|x - T_1(x)\|^2, \\ \|T_2(x) - z\|^2 &\le \|x - z\|^2 - \nu_2\|x - T_2(x)\|^2.\end{aligned}$$

Multiplying the first inequality by α_1 and the second one by α_2, and summing the left- and right-hand sides, we obtain

$$\begin{aligned}\|T(x) - z\|^2 &\le \alpha_1\|T_1(x) - z\|^2 + \alpha_2\|T_2(x) - z\|^2 \\ &\le \|x - z\|^2 - \nu_1\alpha_1\|x - T_1(x)\|^2 - \nu_2\alpha_2\|x - T_2(x)\|^2 \\ &\le \|x - z\|^2 - \min\{\nu_1, \nu_2\}\, \|T(x) - z\|^2,\end{aligned}$$

i.e., $T = \alpha_1T_1 + \alpha_2T_2 \in \mathcal{P}_M^{\nu}$, where $\nu = \min\{\nu_1, \nu_2\}$.

By induction on m, we complete the statement of the theorem. □

Remark 1.9. The statement of the theorem is valid for any order of indices in the superposition $T = T_{j_1} T_{j_2} \dots T_{j_m}$.

Definition 1.10. An operator (mapping) $T : \mathcal{H} \to \mathcal{H}$ is called *monotone*, if

$$(T(x) - T(v), x - v) \geq 0 \quad \forall\, x, v \in \mathcal{H}.$$

Theorem 1.11. *If $T : \mathcal{D}(T) = \mathcal{H} \to \mathcal{H}$ and $T \in \mathcal{K}$, then the operator $F = I - T$ is monotone, and the following property is valid:*

$$x_j \rightharpoonup x, \quad x_j - T(x_j) \to 0 \Longrightarrow x = T(x), \tag{1.7}$$

where the symbol "$\rightharpoonup$" means weak convergence.

Proof. The monotonicity of this operator follows from the relation

$$(F(x) - F(v), x - v) = \|x - v\|^2 - (T(x) - T(v), x - v) \geq 0.$$

Let now the assumption in (1.7) be satisfied. Then for any element $u \in \mathcal{H}$ we have

$$\lim_{j \to \infty} (F(u) - F(x_j),\, u - x_j) = (F(u), u - x) \geq 0.$$

In particular, for $u_t = x + tw$, where $t > 0$ and w is an arbitrary element from $\mathcal{H}$, we have $t(F(u_t), w) \geq 0$. Cancelling the last expression by t and passing to the limit as $t \to 0$, we obtain $(F(x), w) \geq 0$, from which $F(x) = x - T(x) = 0$, i.e., $x \in \mathrm{Fix}(T)$. □

1.2 Structure of the set of fixed points

Before the formulation of the main result on the existence of fixed points, let us clarify the structure of the set of fixed points for a large class of mappings, namely, for the class $\mathcal{K}_M$.

Lemma 1.12. *If $T \in \mathcal{K}_M$, then the set $M = \mathrm{Fix}(T)$ is closed and convex.*

Proof. Let $x_0, x_1 \in \mathrm{Fix}(T) = M$ and $x_\lambda = (1 - \lambda)x_0 + \lambda x_1$. Then from the lemma conditions, we have

$$\|T(x_\lambda) - x_0\| \leq \|x_\lambda - x_0\|,$$
$$\|T(x_\lambda) - x_1\| \leq \|x_\lambda - x_1\|.$$

On the other side, we have

$$\begin{aligned}\|x_0 - x_1\| &\le \|x_0 - T(x_\lambda)\| + \|T(x_\lambda) - x_1\| \\ &\le \|x_\lambda - x_0\| + \|x_\lambda - x_1\| \\ &= \lambda\|x_0 - x_1\| + (1-\lambda)\|x_0 - x_1\| = \|x_0 - x_1\|,\end{aligned}$$

i.e., $T(x_\lambda) \in [x_0, x_1]$. From the first two relations, it follows that $T(x_\lambda) \in [x_0, x_\lambda]$, $T(x_\lambda) \in [x_\lambda, x_1]$, i.e., $T(x_\lambda) = x_\lambda$.

Let x be a limit point of the set M and $x \notin M$, i.e., $T(x) \neq x$. Then there exists an element $z \in M$, for which $2\|x - z\| < \|T(x) - x\|$. From this

$$\|T(x) - x\| \le \|T(x) - z\| + \|z - x\| \le 2\|x - z\| < \|T(x) - x\|,$$

that is impossible. Therefore, $x \in M$, i.e., the set M is closed. □

Corollary 1.13. *If $T \in \mathcal{F}_M$ or $T \in \mathcal{P}_M^\nu$, then $M = \mathrm{Fix}(T)$ is convex and closed.*

1.3 Existence of fixed points

We now establish existence of the fixed points for nonexpansive operators (Browder, 1967).

Theorem 1.14. *Let $\mathcal{D}$ be a convex closed bounded subset of the Hilbert space $\mathcal{H}$ and T be a nonexpansive operator acting from $\mathcal{D}$ into $\mathcal{D}$. Then the set $\mathrm{Fix}(T)$ is nonempty, convex, and closed.*

Proof. Introduce the notation $V_s(x) = sT(x) + (1-s)v_0$, where $0 < s < 1$ and v_0 is a fixed element from $\mathcal{D}$. Then V_s is a contractive mapping (with constant $s < 1$) that acts from $\mathcal{D}$ into $\mathcal{D}$. Thus, by the known principle of contractive mappings, V_s has a unique fixed point x_s in $\mathcal{D}$. Since $\mathcal{D}$ is closed and bounded, the set $\{x_s : 0 < s < 1\}$ is weakly compact, and therefore there exists a sequence $x_j = x_{s_j}$ that converges weakly to some element $x_0 \in \mathcal{D}$ for $s_j \to 1$, $j \to \infty$. We show that $x_0 \in \mathrm{Fix}(T)$.

If x is some point from $\mathcal{X}$, then

$$\begin{aligned}\|x_j - x\|^2 &= \|(x_j - x_0) + (x_0 - x)\|^2 \\ &= \|x_j - x_0\|^2 + \|x_0 - x\|^2 + 2(x_j - x_0, x_0 - x),\end{aligned} \tag{1.8}$$

where $2(x_j - x_0, x_0 - x) \to 0$, since $x_j \rightharpoonup x_0$.

Since $s_j \to 1$ for $j \to \infty$, this implies

$$\begin{aligned} T(x_j) - x_j &= \{s_j T(x_j) + (1 - s_j)v_0\} - x_j + (1 - s_j)\{T(x_j) - v_0\} \\ &= \{V_{s_j}(x_j) - x_j\} + (1 - s_j)\{T(x_j) - v_0\} \\ &= (1 - s_j)\{T(x_j) - v_0\} \to 0, \end{aligned} \tag{1.9}$$

using the boundedness of $\{T(x_j)\} \subseteq \mathfrak{D}$.

Substituting $x = T(x_0)$ into (1.8) and passing to the limit, we obtain

$$\lim_{j\to\infty} \{\|x_j - T(x_0)\|^2 - \|x_j - x_0\|^2\} = \|x_0 - T(x_0)\|^2. \tag{1.10}$$

Taking into account that T is a nonexpansive mapping, we obtain the following inequalities:

$$\begin{aligned} \|x_j - T(x_0)\| &\le \|x_j - T(x_j)\| + \|T(x_j) - T(x_0)\| \\ &\le \|x_j - T(x_j)\| + \|x_j - x_0\|, \end{aligned}$$

from which, keeping in mind (1.9), we find

$$\limsup_{j\to\infty} (\|x_j - T(x_0)\| - \|x_j - x_0\|) \le 0.$$

The latter relation implies the inequality

$$\limsup_{j\to\infty} (\|x_j - T(x_0)\|^2 - \|x_j - x_0\|^2) \le 0,$$

which together with (1.10) gives $\|x_0 - T(x_0)\| = 0$, i.e., x_0 is a fixed point of the mapping T in $\mathfrak{D}$.

Convexity and closure of the set $\mathrm{Fix}(T)$ follow from Lemma 1.12. □

2 Strong convergence of iterations for quasi-nonexpansive operators

For the approximation of fixed points of nonlinear mappings T, i.e., for the solution of the equation

$$x = T(x), \tag{2.1}$$

the following two main iterative processes will be studied: *the method of successive approximations*

$$x^{k+1} = T(x^k), \quad k = 0, 1, \ldots \tag{2.2}$$

and the process

$$x^{k+1} = \lambda T(x^k) + (1-\lambda)x^k, \quad k = 0, 1, \ldots, \tag{2.3}$$

which is in fact the method of successive approximations for the operator $T_\lambda = \lambda T + (1-\lambda)I$.

2.1 The case of demi-compact operators

To prove strong convergence, several conditions for the operator T will be needed additionally to the property of affiliation of T to the class $\mathcal{K}$.

Definition 2.1. A mapping $T : \mathcal{X} \to \mathcal{X}$ is called *demi-compact* on $\mathcal{D}$ if it has the following property: existence of a strongly converging subsequence $x_{n_k} \to x$ follows from the fact that $\{x_n\}$ is a bounded subsequence and $T(x_n) - x_n \to 0$.

Theorem 2.2. *Let $\mathcal{H}$ be a Hilbert space, and let an operator $T : \mathcal{H} \to \mathcal{H}$ from the class $\mathcal{K}$ map the convex closed bounded set $\mathcal{D}$ into itself, and in addition let T be demi-compact. Then the set* Fix(T) *of fixed points of T in $\mathcal{D}$ is nonempty, and for any $x^0 \in \mathcal{D}$, the sequence x^k defined by process* (2.3) *converges strongly to $x \in$* Fix$(T) \subset \mathcal{D}$.

Proof. By Theorem 1.14, the set $\mathrm{Fix}(T) \neq \varnothing$ is convex and closed. By Lemma 1.7, the operator $T_\lambda = \lambda T + (1-\lambda)I$ belongs to the class $\mathcal{P}^\nu$ for $\nu = \lambda / (1-\lambda)$; therefore, relation (1.6) holds. Substituting $x = x^k$, $v \in \mathrm{Fix}(T) = \mathrm{Fix}(T_\lambda)$ in this operator, we obtain the following relation:

$$\|x^{k+1} - v\|^2 \le \|x^k - v\|^2 - \lambda / (1-\lambda) \, \|x^k - T_\lambda(x^k)\|^2,$$

from which the boundedness of x^k and the strong convergence follow:

$$\lim_{k\to\infty} \|x^k - T_\lambda(x^k)\| = \frac{1-\lambda}{\lambda} \lim_{k\to\infty} \{\|x^k - v\|^2 - \|x^{k+1} - v\|^2\} = 0.$$

Together with the demi-compactness of the operator T and $T \in \mathcal{K}$, this implies $x^{k_j} \to x$, $T(x^{k_j}) \to T(x) = x$. Thus, convergence of the sequence x^k to a fixed point x follows from the monotonicity, i.e., $\|x^{k+1} - x\| < \|x^k - x\|$. □

Corollary 2.3. *If an operator T can be represented in the form $T = S + U$, where U is a compact operator (see Definition 2.7), and S is such that $(I - S)^{-1}$ exists and is continuous on the domain of values $R(I - S)$, then T is demi-compact.*

Proof. In fact, let x_n be a bounded sequence such that $T(x_n) - x_n \to 0$. By virtue of the compactness, there exists a subsequence x_{n_i}, for which $U(x_{n_i}) \to v \in \mathcal{X}$. Then

$$w_{n_i} = (I - S)x_{n_i} = x_{n_i} - T(x_{n_i}) + U(x_{n_i})$$

converges strongly. Since the operator $(I - S)^{-1}$ is continuous, then $x_{n_k} = (I - S)^{-1} w_n$ also converges strongly. The demi-compactness of T is proved. □

Remark 2.4. If S is a linear bounded operator with the norm $\|S\| < 1$, then by the Banach theorem about the inverse operator, the operator $(I - S)^{-1}$ exists and is bounded; therefore, S satisfies the conditions of Corollary 2.3.

2.2 The case of a linear operator

The assumption on demi-compactness of the operator T in Theorem 2.2 on convergence is a rather burdensome condition, which, as a rule, is not satisfied for ill-posed problems. The following theorem shows that this condition is not needed for a linear operator.

Definition 2.5. A linear normed space $\mathcal{X}$ is called *uniformly convex*, if the following property holds:

$$\forall \varepsilon > 0 \quad \exists \delta(\varepsilon) > 0 : \forall x, v \in \mathcal{X}, \quad \|x\| = \|v\| = 1,$$

$$\|(x + v) / 2\| > 1 - \delta(\varepsilon) \Longrightarrow \|x - v\| < \varepsilon.$$

As it is known, the spaces L_p of integrable with the p-power functions and the Sobolev spaces W_p^n for $1 < p < \infty$ are uniformly convex.

Theorem 2.6. *Let $\mathcal{X}$ be uniformly convex (e.g., a Hilbert space), and let $T : \mathcal{X} \to \mathcal{X}$ be a linear operator from the class $\mathcal{K}$. Then the sequence x^k defined by process* (2.3) *converges strongly to a fixed point of the operator T.*

Proof. We restrict ourselves to the case of a Hilbert space (the general case can be found in (Dotson, 1970)). Since $T \in \mathcal{K}$, the operator $T_\lambda = \lambda T + (1-\lambda)I \in \mathcal{P}_M^\nu$ (as in Lemma 1.7), where $M = \text{Fix}(T) = \text{Fix}(T_\lambda) \neq \varnothing$, because the zero element 0 satisfies $0 \in M$ by virtue of the linearity of T. From that it follows that the inequality

$$\|T_\lambda x - z\|^2 \leq \|x - z\|^2 - \nu \|x - T_\lambda x\|^2$$

is satisfied for any $x \in \mathcal{X}$, $z \in M$. For $x = T_\lambda^n v$ and from this inequality the following relation results:

$$T T_\lambda^n v - T_\lambda^n v \to 0 \quad \forall v \in \mathcal{X}.$$

Then, by linearity and continuity of T, we have

$$T^2 T_\lambda^k v - T_\lambda^k v = T(T T_\lambda^k v - T_\lambda^k v) + (T T_\lambda^k v - T_\lambda^k v) \to T(0) + 0 = 0.$$

By induction we obtain

$$T^m T_\lambda v - T_\lambda v \to 0, \quad n \to \infty, \quad \forall x \in \mathcal{X}, \quad m = 0, 1, \dots .$$

From the linearity, the representation

$$T_\lambda^k = [\lambda T + (1-\lambda) I]^k = \sum_{j=0}^{k} C_j^k (1-\lambda)^{k-j} \lambda^j T^j$$

follows, i.e., T_λ^k is a linear combination of $I, T, T^2, \dots, T^k$, and

$$\|T_\lambda^k\| \le \sum_{j=0}^{k} C_j^k (1-\lambda)^{k-j} \lambda^j \, \|T\|^j \le 1 \quad \forall n.$$

Moreover, since T_λ^k is a polynomial in the operator T, we obtain $T^m T_\lambda^k = T_\lambda^k T^m$, and, therefore, $T_\lambda^k T^m v - T_\lambda^k v \to 0$ as $n \to \infty$ for any $v \in \mathcal{X}$.

Keeping in mind that $\{T_\lambda^k x^0\}$ belongs to a weakly compact set $\{v : \|v\| \le \|x^0\|\}$, this sequence has a weak limit point z. Then from the Eberlein ergodic theorem (Eberlein, 1949), it follows that $x^{k+1} = T_\lambda^k x^0 \to z$ and $z \in \mathrm{Fix}(T)$; and moreover, $T^m z = z$, $m = 0, 1, 2, \dots$. □

2.3 Iterative processes for linear equations of the first kind

In previous subsections the problem of iterative approximation of solutions for equations of the second kind (2.1) was considered. An operator equation of the first kind

$$Ax = y \tag{2.4}$$

can be formally written in the form of the second kind equation

$$x = x - (Ax - y) \equiv T(x) \tag{2.5}$$

and, further, the following iterative process can be studied:

$$x^{k+1} = x^k - (Ax^k - y), \tag{2.6}$$

which we call the *method of a simple iteration* for equation (2.4).

Definition 2.7. An operator (in general a nonlinear one) acting on a pair of Banach spaces $\mathcal{X}$, $\mathcal{Y}$ is called *compact* if it transforms each bounded set into a precompact one (on the notion of compactness, see, for instance, (Kolmogorov and Fomin, 1976)).

Definition 2.8. Let M be a set of solutions for some problem and v^0 be an element of a Hilbert space $\mathcal{X}$. Then the element $\hat{x}$ satisfying the condition

$$\|\hat{x} - v^0\| = \inf\{\|z - v^0\| : z \in M\}$$

is called the v^0-*normal solution* of this problem.

If v^0 is a zero element, then we use the term *normal solution*. In this case we mean a solution with the minimal norm.

Theorem 2.9. *Let $A : \mathcal{H} \to \mathcal{H}$ be a selfadjoint positive semidefinite (i.e., $(Ax, x) \geq 0 \; \forall x \in \mathcal{H}$) compact operator, $\|A\| \leq 1$, and equation* (2.4) *be solvable. Then for any initial approximation $x^0 \in \mathcal{H}$, the iterative process* (2.6) *converges to the v^0-normal solution $\hat{x}$ of equation* (2.4).

Proof. By the Hilbert–Schmidt theorem for a linear self-adjoint compact operator A, there exists an orthonormal basis $\{e_i\}$ of this separable Hilbert space $\mathcal{H}$ which consists of eigenvectors of this operator. Since the operator A is positive semidefinite, its eigenvalues λ_i are nonnegative, and in addition $0 \leq \lambda_i \leq 1$. Let us assume them to be ordered increasingly.

Expand x^0 and y by the basis $\{e_i\}$. Taking into account that $Ae_i = \lambda_i e_i$, we have the representation

$$\begin{aligned} x^{k+1} &= (I - A)x^k + y = (I - A)^2 x^{k-1} + (I - A)y + y \\ &= (I - A)^{k+1} x^0 + \sum_{j=0}^{k} (I - A)^j y \qquad (2.7) \\ &= \sum_{i=1}^{\infty} (1 - \lambda_i)^{k+1} x_i^0 e_i + \sum_{i=1}^{\infty} \Big[\sum_{j=0}^{k} (1 - \lambda_i)^j \Big] y_i e_i. \end{aligned}$$

We next find an expression for a solution $\bar{x}$ of equation (2.4). Substituting the expansion

$$x = \sum_{i=1}^{\infty} \bar{x}_i e_i,$$

in this equation, we obtain

$$Ax = \sum_{i=1}^{\infty} \bar{x}_i Ae_i = \sum_{i=1}^{\infty} \lambda_i \bar{x}_i e_i = \sum_{i=1}^{\infty} y_i e_i,$$

from which we find $\bar{x}_i = y_i / \lambda_i$ if $x_i \neq 0$; here, $\bar{x}_i$ is arbitrary if $\lambda_i = 0$, and then $y_i = 0$.

Now representation (2.7) can be written in the form

$$\begin{aligned} x^{k+1} = \sum_{i \in I_0} (1-\lambda_i)^{k+1} x_i^0 e_i + \sum_{i \in I_1} (1-\lambda_i)^{k+1} x_i^0 e_i \\ + \sum_{i \in I_1} \frac{1-(1-\lambda_i)^{k+1}}{\lambda_i} y_i e_i, \end{aligned} \tag{2.8}$$

where $I_0 = \{i : \lambda_i = 0\}$, $I_1 = \{i : \lambda_i \neq 0\}$.

Let us show now that for $k \to \infty$, the following convergence holds:

$$x^{k+1} \to \hat{x} = \sum_{i \in I_0} x_i^0 e_i + \sum_{i \in I_1} \frac{y_i}{\lambda_i} e_i.$$

In fact, the second term in relation (2.8) tends to zero, since in the representation

$$\sum_{i \in I_1} (1-\lambda_i)^{k+1} x_i^0 e_i = \sum_{i \in I_1, i \leq N} (1-\lambda_i)^{k+1} x_i^0 e_i + \sum_{i \in I_1, i > N} (1-\lambda_i)^{k+1} x_0^0 e_i$$

the first sum tends to zero by virtue of $0 < \lambda_i < 1$, and the second sum can be made arbitrary small for sufficiently large N, since the following estimate holds:

$$\Big\| \sum_{i \in I_1, i > N} (1-\lambda_i)^{k+1} x_i^0 e_i \Big\|^2 \leq \sum_{i=N+1}^{\infty} |x_i^0|^2.$$

Using an analogous reasoning, we see that, as $k \to \infty$,

$$\sum_{i \in I_1} \frac{1-(1-\lambda_i)^{k+1}}{\lambda_i} y_i e_i \longrightarrow \sum_{i \in I_1} \frac{y_i}{\lambda_i} e_i.$$

It is evident that $\hat{x}$ is a solution of equation (2.4). Take an arbitrary solution $\bar{x}$ and decompose the difference as follows,

$$\bar{x} - x^0 = \sum_{i \in I_0} (\bar{x}_i - x_i^0) e_i + \sum_{i \in I_1} \Big(\frac{y_i}{\lambda_i} - x_i^0 \Big) e_i.$$

On the set of all solutions $\bar{x}$, the value

$$\|\bar{x} - x^0\|^2 = \sum_{i \in I_0} (\bar{x}_i - x_i^0)^2 + \sum_{i \in I_1} \Big(\frac{y_i}{\lambda_i} - x_i^0\Big)^2$$

achieves the minimal magnitude for $\bar{x}_i = x_i^0$, $i \in I_0$. Since

$$\hat{x} - x^0 = \sum_{i \in I_1} \Big(\frac{y_i}{\lambda_i} - x_i^0\Big) e_i,$$

this means that the solution $\bar{x}$, where the minimum of $\|\bar{x} - x^0\|^2$ is achieved, coincides with $\hat{x}$.

Thus, the sequence of iterations x^k of process (2.6) converges to a solution that has minimal deviation from x^0. □

Remark 2.10. Since in this theorem $\|A\| \le 1$, it holds $0 \le \lambda_i \le 1$; therefore, $\|I - A\| = \sup\{1 - \lambda_i : 0 \le \lambda_i \le 1\} \le 1$, so the operator T in (2.5) belongs to the class $\mathcal{K}$.

Remark 2.11. In the theorem, the selfadjointness and nonnegativity of the operator A is assumed. To avoid this assumption, it is possible to pass preliminarily from equation (2.4) to the equation

$$A^* A x = A^* y,$$

which, in the case of solvability of equation (2.4), is equivalent to the original equation and has the required properties.

The linear equation (2.4) can be formally represented in the following form:

$$Ax + \alpha x = \alpha x + y \tag{2.9}$$

with the parameter $\alpha > 0$. For a selfadjoint, positive semidefinite operator A, the operator $A + \alpha I$ has a bounded inverse operator; this allows one to write (2.9) in the form of the equivalent operator equation of the second kind

$$x = (A + \alpha I)^{-1}(\alpha x + y) \equiv T(x) \tag{2.10}$$

with the operator $T \in \mathcal{K}$, since $\|(A + \alpha I)^{-1}\alpha\| = \sup\{\alpha / (\lambda + \alpha) : \lambda \in \sigma(A)\} \le 1$ ($\sigma(A)$ is the spectrum of the operator A).

Writing the method of successive approximations for equation (2.10), we arrive at the *implicit iterative scheme*

$$x^{k+1} = (A + \alpha I)^{-1}(\alpha x^k + y) \tag{2.11}$$

for the original equation (2.4).

Theorem 2.12. *Let the assumptions of Theorem 2.9 be satisfied without the requirement* $\|A\| \leq 1$. *Then for process* (2.11), *the conclusion of Theorem 2.9 on the strong convergence of iterations to* x^0, *i.e., to the normal solution of equation* (2.4), *is valid.*

Proof. The proof is implemented by the method from the previous theorem. □

Remark 2.13. The statements of Theorems 2.9 and 2.12 are valid for any arbitrary linear bounded (not necessarily compact) operator; see (Vainikko, 1980).

3 Properties of iterations for the pseudo-contractive mappings

3.1 The main theorem on convergence of iterations

In Subsections 2.1 and 2.2 of the previous section, the step operator T_λ in the process (2.3) belongs to the class $\mathcal{P}_M^\nu \subset \mathcal{P}^\nu$ as a convex combination of nonexpansive and identity operators. So, actually (under some additional assumptions), the strong convergence has been proved for the method of successive approximations for the operator $T_\lambda \in \mathcal{P}_M^\nu$. It is natural to consider the general situation: the method of successive approximations for an arbitrary pseudo-contractive operator T, i.e., $T \in \mathcal{P}_M^\nu$. It reveals that in this case, it is possible, generally speaking, to obtain weak convergence only (Martinet, 1972; Vasin, 1988).

Theorem 3.1. *Let an operator* $T : \mathcal{H} \to \mathcal{H}$ *be from the class* $\mathcal{P}_M^\nu$ *and satisfy the relation*

$$x_j \rightharpoonup x, \quad x_j - T(x_j) \to 0 \Longrightarrow x \in \mathrm{Fix}(T). \tag{3.1}$$

Then for the iteration process (2.2) *the following properties are valid:*

1) $x^k \rightharpoonup \hat{x} \in \mathrm{Fix}(T)$;
2) $\inf\{\lim_{k\to\infty} \|x^k - z\| : z \in \mathrm{Fix}(T)\} = \lim \|x^k - \hat{x}\|$;
3) *either* $\|x^{k+1} - \hat{x}\| < \|x^k - \hat{x}\|$ *for any* k, *or the sequence* x^k *is stationary beginning from some* $k_0 \geq 0$, *i.e.,* $x^{k_0} = x^{k_0+1} = \cdots = \hat{x}$;
4) *the estimate is valid*

$$\sum_{k=0}^{\infty} \|x^{k+1} - x^k\|^2 \leq \|x^0 - z\|^2 / \nu \quad \forall\, z \in \mathrm{Fix}(T).$$

Proof. Since $T \in \mathcal{P}_M^\nu$ (see Definition 1.3), the substitution of $x = x^k$ into inequality (1.3) gives the relation

$$\|x^{k+1} - z\|^2 \le \|x^k - z\|^2 - \nu \, \|T(x^k) - x^k\|^2, \tag{3.2}$$

from which the existence of the limits follows:

$$\lim_{k\to\infty} \|x^k - z\| = d(z) \ge 0 \quad \forall \, z \in \operatorname{Fix}(T), \tag{3.3}$$

$$\lim_{k\to\infty} \|T(x^k) - x^k\| = 0. \tag{3.4}$$

A summation from $k = 0$ to $k = N$ in inequality (3.2) gives the following estimate:

$$\sum_{k=0}^{N} \|x^{k+1} - x^k\|^2 \le \frac{1}{\nu} \sum_{k=0}^{N} (\|x^k - z\| - \|x^{k+1} - z\|^2) \le \frac{\|x^0 - z\|^2}{\nu},$$

which implies property 4).

From (3.3) the boundedness of x^k follows, so there exists a weakly converging subsequence

$$x^{k_j} \rightharpoonup \hat{x}. \tag{3.5}$$

By virtue of condition (3.1), we conclude from (3.4) and (3.5) that $\hat{x} \in \operatorname{Fix}(T)$.

Establish now the uniqueness of the weak limit point, i.e., the convergence of the entire sequence $x^k \rightharpoonup \hat{x}$. Let x_1, x_2 be two weak limit points of the sequence x^k:

$$x^{k_i} \rightharpoonup x_1, \quad x^{k_j} \rightharpoonup x_2, \quad x_1, \, x_2 \in \operatorname{Fix}(T).$$

Then, accordingly to (3.3)

$$\lim_{k\to\infty} \|x^k - x_1\| = d_1, \quad \lim_{k\to\infty} \|x^k - x_2\| = d_2.$$

From the representation

$$\|x^{k_i} - x_2\|^2 = \|x^{k_i} - x_1\|^2 + 2\,(x^{k_i} - x_1, x_1 - x_2) + \|x_1 - x_2\|^2$$

for $i \to \infty$, we find that $d_2^2 - d_1^2 - \|x_1 - x_2\|^2 = 0$. Changing positions of x_1, x_2 and using $x^{k_j} \rightharpoonup x_2$, we obtain $d_1^2 - d_2^2 - \|x_1 - x_2\|^2 = 0$; therefore, $x_1 = x_2$.

Property 2) follows from the identity

$$\|x^k - z\|^2 = \|x^k - \hat{x}\|^2 + 2\,(x^k - \hat{x}, \, \hat{x} - z) + \|\hat{x} - z\|^2,$$

in which it is necessary to pass to the limit with taking into account that $x^k \rightharpoonup \hat{x} \in \operatorname{Fix}(T)$.

Property 3) directly follows from inequality (3.2). □

Corollary 3.2. *Let in the theorem $T \in \mathcal{K}_M$ be used instead of $T \in \mathcal{P}_M^\nu$. Then the statements of this theorem are valid for process* (2.3).

Proof. This follows from the fact that the property $T \in \mathcal{K}_M$ implies $T_\lambda = \lambda T + (1-\lambda)I \in \mathcal{P}_M^\nu$ for $\nu = (1-\lambda)/\lambda$, and condition (3.1) implies the assertions for the operator T_λ. □

Corollary 3.3. *If $T \in \mathcal{P}^\nu$ (or $T \in \mathcal{K}$) on $\mathcal{X}$ and* $\operatorname{Fix}(T) \neq \emptyset$, *then for the iterative scheme* (2.2) *(or* (2.3)*), in particular, for processes* (2.6) *and* (2.11) *the conclusion of Theorem 3.1 is valid.*

Proof. It is sufficient to establish that for the operator T relation (3.1) holds. This fact follows from Theorem 1.11. This completes the proof. □

Definition 3.4. An operator T acting on a pair of linear normed spaces $\mathcal{X}, \mathcal{Y}$ is called *(sequentially) weakly closed*, if from the conditions $x_k \in \mathcal{D}(T)$, $x_k \rightharpoonup x$ and $T(x_k) \rightharpoonup y$ it follows $x \in \mathcal{D}(T)$, $T(x) = y$.

Corollary 3.5. *The statement of Theorem 3.1 remains valid if instead of condition* (3.1) *one requires (sequential) weak closedness of the operator T.*

Proof. Actually, on the basis of the assumption that $T \in \mathcal{P}_M^\nu$, relations (3.4) and (3.5) have been obtained in proving the theorem; namely,

$$x^{k_j} \rightharpoonup \hat{x}, \quad x^k - T(x^k) \to 0,$$

from which it follows that $T(x^{k_j}) \rightharpoonup \hat{x}$. Then by virtue of the weak closedness of T, we conclude that $\hat{x} \in \mathcal{D}(T)$ and $T(\hat{x}) = \hat{x}$.

Proofs of all other properties can be employed without changes. □

Corollary 3.6. *The theorem is valid for a multi-valued operator T. In this case the iterations x^k weakly converge to an element x that satisfies the inclusion $x \in T(x)$ if condition* (3.1) *is changed as follows:*

$$x_j \rightharpoonup x, \quad x_j - y_j \to 0,$$

$$y_j \in T(x_j) \Longrightarrow x \in T(x). \qquad \square$$

Remark 3.7. Under the conditions of Theorem 3.1 and its corollaries, the result, generally speaking, can not be enhanced to obtain strong convergence for iterations. This follows from the work (Genel and Lindenstrauss, 1975) where an example of a set $\mathcal{D}$ in the space l_2, a mapping $T : \mathcal{D} \to \mathcal{D}$ from the class $\mathcal{K}$, and an initial approximation is given, for which process (2.3) weakly converges, but strong convergence does not hold.

Thus, without additional assumptions, Theorem 3.1 and its corollaries do guarantee only weak convergence. But it is possible to show an evident case of strong convergence of the iterations; namely, the following corollary holds.

Corollary 3.8. *If the linear hull* $\operatorname{Lin}\{x^k\}_0^\infty$ *has finite dimension, then the iterative process* (2.2) *converges strongly to a fixed point of the operator* T. □

3.2 Superpositions of pseudo-contractive operators

In some cases the original problem admits the construction of an iterative process where the step operators are a superposition or a convex combination of mappings that are very simple and economical in their implementation. Thus, the problem is divided into subproblems. For each subproblem its own iterative operator is constructed. Further by means of some aggregative operation, the iterative operator for the whole problem is generated by the considered fragments. It is therefore possible to construct a solution operator from the same class as the generating mappings. One application of such techniques for the pseudo-contractive operators follows from the next theorem (Vasin, 1988).

Theorem 3.9. *Let* $M = \bigcap_{i=1}^m M_i \neq \varnothing$, $T_i \in \mathcal{P}_{M_i}^{\nu_i}$, *and for each operator* T_i *relation* (3.1) *be satisfied. Then for the iterative process* (2.2) *where* $T = T_m T_{m-1} \ldots T_1$ *or* $T = \sum_{i=1}^m \lambda_i T_i$, $0 < \lambda_i < 1$, $\sum \lambda_i = 1$, *the conclusion of Theorem 3.1 is valid.*

Proof. By Theorem 1.8 the operator $T = T_m T_{m-1} \ldots T_1 \in \mathcal{P}_M^\nu$, where $\nu = \min_{1\le i\le m}\{\nu_i\} / 2^{m-1}$, or $T = \sum_{i=1}^n \lambda_i T_i \in \mathcal{P}_M^\nu$, where $\nu = \min_{1\le i\le m}\{\nu_i\}$. Thus, to use the reasonings of Theorem 3.1, it is necessary to be sure that property (3.1) holds at least for the iterative sequence x^k. It is sufficient to consider only the case of $m = 2$.

Let $z \in M = M_1 \cap M_2$, $T = T_2 T_1$, then

$$\begin{aligned}\|T_2T_1(x) - z\|^2 &\le \|T_1(x) - z\|^2 - \nu_2\|(I - T_2)T_1(x)\|^2 \\ &\le \|x - z\|^2 - \nu_1\|(I - T_1)x\|^2 - \nu_2\|(I - T_2)T_1(x)\|^2 \\ &\le \|x - z\|^2 - \nu\|x - T_2(T_1(x))\|^2.\end{aligned}$$

If now $x^{k_i} \rightharpoonup x$, $x^{k_i} - T(x^{k_i}) \to 0$, then from the previous collection of inequalities for $x = x^{k_i}$, we obtain the following relations:

$$x^{k_i} - T_1(x^{k_i}) \to 0, \quad z_{k_i} = T_1(x^{k_i}) \rightharpoonup x, \quad z_{k_i} - T_2(z_{k_i}) \to 0,$$

from which, in correspondence with the theorem assumptions that for T_i condition (3.1) is satisfied, we obtain

$$x \in \operatorname{Fix}(T_1) \cap \operatorname{Fix}(T_2) = M_1 \cap M_2 = M,$$

i.e., the claimed relation (3.1) holds for the operator $T = T_2 T_1$.

If $T = \lambda_1 T_1 + \lambda_2 T_2$, then, since in Theorem 1.8 the following chain of inequalities was established,

$$\begin{aligned} \|T(x) - z\|^2 &\le \|x - z\|^2 - \nu_1 \lambda_1 \|x - T_1(x)\|^2 - \nu_2 \lambda_2 \|x - T_2(x)\|^2 \\ &\le \|x - z\|^2 - \min\{\nu_1, \nu_2\} \|T(x) - x\|^2, \end{aligned}$$

the verification of property (3.1) is carried out in a similar way as for the operator $T = T_2 T_1$. □

3.3 Examples of the pseudo-contractive operators

Consider several examples.

1. Let Q be a convex closed subset of a Hilbert space $\mathcal{H}$ and f be a functional which is convex and weakly semi-continuous from below. Define the mapping

$$\mathrm{S}_Q^f : v \to \arg\min\{f(x) + (1/2)\|x - v\|^2 : x \in Q\},$$

which is called the *prox-mapping*. It is easy to check that the problem considered in the definition of S_Q^f is solvable for any $v \in \mathcal{H}$; thus, the definition S_Q^f is correct.

Lemma 3.10. *The prox-mapping* S_Q^f *belongs to the class* $\mathcal{P}^1$, *i.e., the following relation holds for it:*

$$\begin{aligned} \|T(x) - T(v)\|^2 \le \|x - v\|^2 - \|T(x) - x - (T(v) - v)\|^2 \\ \forall x, v \in \mathcal{H}. \end{aligned} \tag{3.6}$$

Moreover, the set $\operatorname{Fix}(\mathrm{S}_Q^f)$ *of fixed points coincides with the set* M *of solutions for the problem (in the case* $M \neq \varnothing$*)*

$$\min\{f(x) : x \in Q\}.$$

Proof. The proof of this fact can be found in the monograph (Vasin and Ageev, 1995). □

Corollary 3.11. $S_Q^f \in \mathcal{P}_M^1$ *(see Definition 1.3).* □

2. Consider the *operator* P_Q of *metric projection* (or, shortly, *metric projection*) onto a closed convex set $Q \in \mathcal{H}$,

$$P_Q(x) = \arg\min\{\|x - v\|^2 : v \in Q\}. \tag{3.7}$$

It is evident that P_Q is a particular case of the prox-mapping S_Q^f for $f(x) \equiv 0$, so we obtain the following corollary from Lemma 3.10.

Corollary 3.12. *The metric projection P_Q onto an arbitrary convex closed subset Q of a Hilbert space $\mathcal{H}$ belongs to the class $\mathcal{P}^1$. It is therefore nonexpansive, Q-pseudo-contractive, and a Fejér operator (see Definitions 1.2–1.4).*

3. Let Q be a convex closed subset of a Hilbert space $\mathcal{H}$. The *metric projection with relaxation* is given by the formula

$$P_Q^\lambda(x) = x - \lambda(x - \bar{x}), \tag{3.8}$$

where λ is the *coefficient of relaxation*, $0 < \lambda < 2$, and $\bar{x} = P_Q(x)$ is the metric projection defined by (3.7). Introduce the *functional of distance* $\rho_Q(x)$ from the point x to the set Q. The functional $\rho_Q(x)$ is differentiable at points $x \notin M$; for this,

$$\begin{aligned} \nabla\rho_Q(x) &= (x - \bar{x}) / \|x - \bar{x}\|, \\ (\nabla\rho_Q(x), z - x) &\le \rho_Q(z) - \rho_Q(x). \end{aligned} \tag{3.9}$$

Moreover, from the definition of $P_Q^\lambda(x)$, it follows:

$$\rho_Q(x) = \|x - \bar{x}\|^2 = \|P_Q^\lambda(x) - x\|^2 / \lambda^2 \tag{3.10}$$

and $\mathrm{Fix}(P_Q^\lambda) = Q$. Note also that the functional is nondifferentiable at the boundary of the set Q, but on the other hand it has a subdifferential at any point of the space $\mathcal{H}$ (see Dem'yanov and Vasil'ev, 1981; Clark, 1988).

Lemma 3.13. *The inequality*

$$\|P_Q^\lambda(x) - z\|^2 \le \|x - z\|^2 - \frac{2-\lambda}{\lambda}\|P_Q^\lambda(x) - x\|^2 \tag{3.11}$$

holds for any $x \in \mathcal{H}$, $z \in Q$, i.e., P_Q^λ belongs to the class $\mathcal{P}_Q^\nu$ for $\nu = (2-\lambda)/\lambda$.

Proof. Taking into account relations (3.9) and (3.10), we have for $z \in Q$ and $x \notin Q$

$$\begin{aligned}
&\|P_Q^\lambda(x) - z\|^2 \\
&\quad = \|x - z\|^2 + \lambda^2 \|x - \bar{x}\|^2 + 2\lambda \left(\frac{x - \bar{x}}{\|x - \bar{x}\|}, z - x \right) \|x - \bar{x}\| \\
&\quad \le \|x - z\|^2 + \lambda^2 \|x - \bar{x}\|^2 + 2\lambda \left(\|z - \bar{z}\| - \|x - \bar{x}\| \right) \|x - \bar{x}\| \\
&\quad = \|x - z\|^2 - \lambda (2 - \lambda) \|x - z\|^2 \\
&\quad = \|x - z\|^2 - \frac{(2 - \lambda)}{\lambda} \|P_Q^\lambda(x) - x\|^2.
\end{aligned}$$

Here, $\bar{z} = P_Q(z)$, $x \notin Q$. For $x \in Q$, relation (3.11) obviously holds. □

4. Consider now the mapping of the form

$$T(x) = \begin{cases} x - \lambda \dfrac{d(x)\, e(x)}{\|e(x)\|^2}, & d(x) > 0, \\ x, & d(x) \le 0, \end{cases} \tag{3.12}$$

where $d(x)$ is a convex subdifferentiable functional given on a Hilbert space $\mathcal{H}$, and $e(x)$ is its subgradient. Define the set $M = \{x : d(x) \le 0\} \ne \varnothing$ so that $e(x) \ne 0$ for any element $x \notin M$.

Lemma 3.14. *For* $0 < \lambda < 2$ *the operator* T *defined by relation* (3.12) *belongs to the class* $\mathcal{P}_M^\nu$, *where* $\nu = (2 - \lambda) / \lambda$, $0 < \lambda < 2$.

Proof. Let $z \in M$, $x \notin M$. Then, taking into account that $d(x) > 0$, $d(z) \le 0$ and the fact that $e(x)$ is the subgradient of the functional $d(x)$, we have the estimate

$$\begin{aligned}
\|T(x) - z\|^2 &= \|x - z\|^2 - 2\lambda \frac{d(x)}{\|e(x)\|^2} (e(x), x - z) + \lambda^2 \frac{d^2(x)}{\|e(x)\|^2} \\
&\le \|x - z\|^2 - 2\lambda \frac{d(x)}{\|e(x)\|^2} (d(x) - d(z)) + \lambda^2 \frac{d^2(x)}{\|e(x)\|^2} \\
&= \|x - z\|^2 - (2 - \lambda)\lambda \frac{d^2(x)}{\|e(x)\|^2} + 2\lambda \frac{d(x)\, d(z)}{\|e(x)\|^2} \\
&\le \|x - z\|^2 - \frac{2 - \lambda}{\lambda} \|x - T(x)\|^2.
\end{aligned}$$

□

Remark 3.15. In the four examples shown above, it was established that all the mappings belong to the class $\mathcal{P}_M^{\nu}$. Therefore, under the additional assumption that property (3.1) is satisfied, we are in the situation of Theorem 3.1 that guarantees weak convergence of the iterations to a fixed point. In particular, for the operator T of form (3.12) and for a choice of $d(x)$ in the form

$$d(x) = \sum_{j=1}^{m} k_j \, [f_j^+(x)]^{\mu}, \quad k_j > 0, \;\; \mu \geq 1, \tag{3.13}$$

where f_j are convex functionals that are semi-continuous from below and for which the subdifferentials $\partial f_j(x)$ are bounded mappings (i.e., the image of a bounded set is bounded), the mentioned property (3.1) holds. By the way, the property of boundedness of the subdifferential is evidently satisfied for the quadratic functional $f_j(x) = \|A_j x - y\|^2$ (A_j is a linear operator) and affine functionals $f_j(x) = (x, a_j) - b_j$.

In the next lemma, a sufficient condition for boundedness of the subdifferential $\partial f(x)$ is given as for the mapping $\partial f : \mathcal{H} \to 2^{\mathcal{H}}$.

Lemma 3.16. *Let a convex functional* $f : \mathcal{H} \to \mathbb{R}$ *be subdifferentiable on* $\mathcal{H}$ *and bounded (i.e., the image of each bounded set is bounded). Then the subdifferential* $\partial f(x)$ *is a bounded mapping, i.e., for any bounded set* $M \subset \mathcal{H}$ *we have*

$$\sup \{\|h\| : h \in \partial f(x), \, x \in M\} \leq C(M) < \infty. \tag{3.14}$$

Proof. Let M be a bounded subset of the space $\mathcal{H}$ and $x \in M$. By definition of the *subgradient* h *of functional* $f(x)$ at the point x, the following relation holds:

$$(h, v - x) \leq f(v) - f(x) \quad \forall \, v \in \mathcal{H}.$$

Taking $v = x + \varepsilon \frac{h}{\|h\|}$, where $h \in \partial f(x)$, $\varepsilon > 0$, we obtain the inequality

$$\|h\| \leq \frac{1}{\varepsilon} \left\{ f\left(x + \varepsilon \frac{h}{\|h\|} \right) - f(x) \right\}. \tag{3.15}$$

Since by assumption the functional f is bounded, the right-hand side of inequality (3.15) is bounded on the set M, i.e.,

$$\sup \frac{1}{\varepsilon} \left\{ f\left(x + \varepsilon \frac{h}{\|h\|} \right) - f(x) : x \in M, \, h \in \partial f(x) \right\} \leq C(M) < \infty,$$

that implies estimate (3.14). □

Corollary 3.17. *Let a convex functional $f : \mathcal{H} \to \mathbb{R}$ satisfy a Lipschitz condition, i.e., for some constant $c > 0$*

$$|f(x) - f(v)| \leq c\,\|x - v\| \quad \forall\, x, v \in \mathcal{H}. \tag{3.16}$$

Then the functional f is subdifferentiable on $\mathcal{H}$ and its subdifferential is a bounded mapping.

Proof. Since the Lipschitz property implies continuity, the subdifferentiability follows from the known fact of convex analysis (Ioffe and Tikhomirov, 1974). If M is a bounded set, then it belongs to some sphere $\overline{S}_r(a)$. Then from inequality (3.16), the estimate follows

$$|f(x)| \leq |f(a)| + c\,r,$$

i.e., the boundedness of the range follows. □

In the general case, the subdifferential, in particular, the gradient of a convex functional, is not necessarily a bounded mapping as the following example shows.

5. Consider the convex functional

$$f(x) = \int_0^1 [x'(t)]^2\,dt \tag{3.17}$$

on the subset of functions $D(f) = \{x(t) : x' \in C\,[0,1],\ x(0) = x(1) = 0,\ \int_0^1 [x''(t)]^2\,dt < \infty\}$ of the space $L_2[0,1]$. Taking into account zero boundary conditions, we obtain

$$\lim_{\lambda \to 0} \frac{f(x + \lambda h) - f(x)}{\lambda} = 2 \int_0^1 x'(t)\,h'(t)\,dt = -2 \int_0^1 x''(t)\,h(t)\,dt.$$

Thus, functional (3.17) is differentiable on the set $D(f)$ in the space $L_2[0,1]$, and its gradient is $\nabla f(x) = -2x''(t)$. This mapping acting from $D(f) \subset L_2[0,1]$ into $L_2[0,1]$ is unbounded, since it transfers the bounded set

$$M = \{\sin n\pi t,\ n = 1, 2, \ldots\} \subset L_2[0,1]$$

into an unbounded one, so

$$\sup\left\{ \int_0^1 \|\nabla f(x)\|^2\,dx : x \in M \right\}$$
$$= 4 \sup\left\{ \int_0^1 [x''(t)]^2\,dt : x \in M \right\} = 2 \sup_n (\pi n)^4 = \infty.$$

4 Iterations with the correcting multipliers

From the results of the previous section (see Remark 3.7) it is seen that for nonexpansive and pseudo-contractive mappings T, generally speaking, it is not possible to construct strongly converging algorithms on the basis of the classical iterative schemes (2.2) and (2.3). For obtaining a strong approximation of fixed points, one special modification (the iterative regularization) is needed by means of additional varying parameters (correcting multipliers).

Consider the iterative process (Halperin, 1967)

$$x^{k+1} = \gamma_{k+1} T(x^k) + (1 - \gamma_{k+1})\, v_0, \quad k = 0, 1, \dots, \tag{4.1}$$

which can be regarded as a modification of scheme (2.2) by means of the correcting (damping) multipliers γ_k; here, v_0 is some fixed element from $\mathcal{H}$ and $0 < \gamma_k < 1$. It turns out that for some special choice of γ_k, the process (4.1), in contrast to (2.2), generates a strongly converging iterative sequence for the operator $T \in \mathcal{K}$. It is natural to call such a process (4.1) the *method of correcting multipliers*.

4.1 Stability of fixed points on parameter

Consider the equation

$$x = \gamma\, T(x) + (1 - \gamma)\, v^0 \tag{4.2}$$

with the operator $T \in \mathcal{K}$ and $0 < \gamma < 1$ and show that its solutions approximate some point from $\mathrm{Fix}(T)$ with respect to the strong topology of a Hilbert space $\mathcal{H}$ (Browder, 1967).

Theorem 4.1. *Let $T : D(T) = \mathcal{H} \to \mathcal{H}$ be a nonexpansive mapping and let $\mathcal{D}$ be a bounded, convex, and closed subset of a Hilbert space $\mathcal{H}$, which is mapped into itself by the operator. Then for any element $v_0 \in \mathcal{D}$ and parameter $0 < \gamma < 1$, there exists a unique solution $x_\gamma \in \mathcal{D}$ of equation* (4.2), *and for this, $\lim_{\gamma\to 1} \|x_\gamma - \hat{x}\| = 0$, where $\hat{x}$ is a fixed point of the operator T closest to v_0, i.e., $\hat{x}$ is the v_0-normal solution of equation* (2.1).

Proof. Since the operator $T_\gamma(x) = \gamma T(x) + (1 - \gamma)\, v_0$ satisfies the condition

$$\|T_\gamma(x) - T_\gamma(v)\| \le \gamma\, \|T(x) - T_\gamma(v)\| \le \gamma\, \|x - v\| \quad \forall\, x, v \in \mathcal{H},$$

equation (4.2) has a unique solution (a fixed point of the operator T_γ) $x_\gamma \in \mathcal{D}$ for $0 < \gamma < 1$ on the basis of the Banach principle of fixed points. By virtue of the weak compactness of the set $\mathcal{D}$, there exists a weakly converging subsequence

$$x_{\gamma_j} \rightharpoonup x, \quad \gamma_j \to 1. \tag{4.3}$$

Since $\{T(x_{\gamma_j})\} \subset \mathcal{D}$, this subsequence is bounded, therefore,

$$\begin{aligned} x_{\gamma_j} - T(x_{\gamma_j}) &= x_{\gamma_j} - \{\gamma_j\, T(x_{\gamma_j}) + (1-\gamma_j)\, v_0\} \\ &\quad + (1-\gamma_j)\, v_0 - (1-\gamma_j)\, T(x_{\gamma_j}) \\ &= (1-\gamma_j)\, v_0 - (1-\gamma_j)\, T(x_{\gamma_j}) \to 0 \end{aligned} \tag{4.4}$$

for $\gamma_j \to 1$. From relations (4.3), (4.4), and Theorem 1.11, we conclude that $x \in \mathrm{Fix}(T) \subset \mathcal{D}$.

Let $\hat{x} \in \mathrm{Fix}(T)$ be the fixed point with minimal deviation from v_0. Such a point exists and is unique, since by Theorem 1.14 the set $\mathrm{Fix}(T)$ is not empty and closed. By definition of x_{γ_j} and $\hat{x}$, we have

$$\begin{aligned} (1-\gamma_j)\, x_{\gamma_j} + \gamma_j\, \{x_{\gamma_j} - T(x_{\gamma_j})\} &= (1-\gamma_j)\, v_0, \\ (1-\gamma_j)\, \hat{x} + \gamma_j\, \{\hat{x} - T(\hat{x})\} &= (1-\gamma_j)\, \hat{x}. \end{aligned}$$

Subtracting term-by-term from the first equality the second one and making the scalar product with the remainder $x_{\gamma_j} - \hat{x}$, we obtain

$$\begin{aligned} (1-\gamma_j)\, (x_{\gamma_j} - \hat{x}, x_{\gamma_j} - \hat{x}) + \gamma_j\, (F(x_{\gamma_j}) - F(\hat{x}), x_{\gamma_j} - \hat{x}) \\ = (1-\gamma_j)\, (v_0 - \hat{x}, x_{\gamma_j} - \hat{x}), \end{aligned}$$

where $F = I - T$ is a monotone operator according to Theorem 1.11. Taking into account this fact, we obtain the following inequality from the last relation:

$$(1-\gamma_j)\, \|x_{\gamma_j} - \hat{x}\|^2 \le (1-\gamma_j)\, (v_0 - \hat{x}, x_{\gamma_j} - \hat{x}),$$

which, after cancelling by $(1-\gamma_j)$, can be written in the form

$$\|x_{\gamma_j} - \hat{x}\|^2 \le (v_0 - \hat{x}, x - \hat{x}) + (v_0 - \hat{x}, x_{\gamma_j} - \hat{x}),$$

where x is the point from $\mathrm{Fix}(T)$, for which $x_{\gamma_j} \rightharpoonup x$.

Taking into account that $\hat{x}$ is the projection of the point v_0 onto the set $M = \mathrm{Fix}(T)$, we conclude that the first term in the right-hand side of the inequality is nonpositive. Moreover, the second term tends to zero, since $x_{\gamma_j} \rightharpoonup x$; so

$$\lim_{\gamma_j \to 1} \|x_{\gamma_j} - \hat{x}\| = 0.$$

Since $\hat{x}$ is the unique limit point, the whole sequence x_γ converges to $\hat{x}$. □

Remark 4.2. It is possible to weaken the condition of the theorem and to require nonexpansiveness of the operator T on the set $\mathcal{D}$ only. In this case in proving the membership of the weak limit point x to the set $\mathrm{Fix}(T)$, it is necessary to use the reasoning from Theorem 1.14 instead of Theorem 1.11. Moreover, in Section 5 of Chapter IV (see Theorems 5.6 and 5.9), analogues of Theorem 4.1 are given about stability on parameter of solutions x_γ of equation (4.2) for operators T from other classes, namely, $\mathcal{F}_M$, $\mathcal{K}_M$.

Remark 4.3. Theorem 4.1 allows one to formulate a two-step algorithm for iterative approximation of the v_0-normal solution $\hat{x}$ of equation (2.1) with the operator $T \in \mathcal{K}$. Actually, take the accuracy $\varepsilon > 0$ of approximation and, on the basis of Theorem 4.1, choose an appropriate $\bar{\gamma}$ in such a way that the estimate $\|\hat{x} - x_{\bar{\gamma}}\| < \varepsilon / 2$ is valid for the solution $x_{\bar{\gamma}}$ of equation (4.2). Further, for fixed $\bar{\gamma}$, using the method of successive approximations

$$\bar{x}^{k+1} = \bar{\gamma} T(\bar{x}^k) + (1 - \bar{\gamma})\, v_0,$$

for sufficiently large $k \geq N(\bar{\gamma})$ we find $\bar{x}^k$ such that $\|x_{\bar{\gamma}} - \bar{x}^k\| < \varepsilon / 2$. Then the final estimate is

$$\|\hat{x} - \bar{x}^k\| \leq \|\hat{x} - x_{\bar{\gamma}}\| + \|x_{\bar{\gamma}} - \bar{x}^k\| < \varepsilon.$$

4.2 One-step iterative process

No doubts, the iterative process (4.1) would be more convenient in applications if there exists a possibility to prescribe *a priori* a sequence of parameters γ_k in such a way that the sequence of iterations x^k of process (4.1) converges strongly to the v_0-normal solution of equation (2.1). It turns out that under some conditions on γ_k this is possible (Halperin, 1967).

Definition 4.4. A numerical sequence γ_i is called *admissible*, if the following conditions are satisfied:

1) $0 < \gamma_i < 1, \ i = 1, 2, \dots;$
2) $\gamma_i < \gamma_{i+1}, \ i = 1, 2, \dots;$
3) $\lim\limits_{i \to \infty} \gamma_i = 1;$
4) there exists a sequence of numbers $n(i)$ such that $n(i+1) > n(i)$, $i = 1, 2, \dots;$
5) $\lim\limits_{i \to \infty} \varepsilon_{i+n(i)} \cdot \varepsilon_i^{-1} = 1$, where $\varepsilon_i = 1 - \gamma_i;$
6) $\lim\limits_{i \to \infty} n(i) \cdot \varepsilon_i = \infty.$

Theorem 4.5. *Let the assumptions of the Theorem 4.1 be satisfied. Then for any initial approximation* $x^0 \in \mathcal{D}$ *and any admissible sequence of* γ_k*, the iterations* (4.1) *converge strongly to* $\hat{x} \in \mathrm{Fix}(T)$*, where* $\hat{x}$ *is the fixed point closest to* v_0.

Proof. Taking into consideration that $\mathcal{D}$ is a bounded set, it is possible to assume that $\mathcal{D} \subset S_r(0)$, where $S_r(0)$ is a ball of radius $r > 0$ with the center at the zero element. Let

$$x_{\gamma_i} = \gamma_i \, T(x_{\gamma_i}) + (1 - \gamma_i) \, v_0,$$

$$x^{i+1} = \gamma_{i+1} \, T(x^i) + (1 - \gamma_{i+1}) \, v_0, \quad i = 0, 1, \dots .$$

Then re-denoting $x_{\gamma_i} = x_i$, we have the estimate

$$\begin{aligned}
&\|x^{l+1} - x_i\| \\
&\quad \le \|\gamma_{l+1} T(x^l) - \gamma_i T(x_i)\| + |\gamma_{l+1} - \gamma_i| \, \|v_0\| \\
&\quad \le \gamma_i \, \|T(x^l) - T(x_i)\| + |\gamma_{l+1} - \gamma_i| \, \|T(x^l)\| + |\gamma_{l+1} - \gamma_i| \, r \\
&\quad \le \gamma_i \, \|x^l - x_i\| + 2r \, |\gamma_n - \gamma_i| \\
&\quad \le \gamma_i^j \, \|x^{l-j+1} - x_i\| + 2r \, |\gamma_n - \gamma_i| \sum_{\nu=0}^{j-1} \gamma_i^\nu \\
&\quad \le \gamma_i^j \, \|x^{l-j+1} - x_i\| + 2r \, |\gamma_n - \gamma_i| \cdot (1 - \gamma_i)^{-1}
\end{aligned}$$

for any $n > l$. From this for $l = k - 1$, $j = k - m$, $m < k$, we obtain

$$\|x^k - x_i\| \le \gamma_i^{k-m} \, \|x^m - x_i\| + 2r \, |\gamma_k - \gamma_i| \, (1 - \gamma_i)^{-1}. \tag{4.5}$$

In the turn for $m = i$, $k = i + n(i)$, the latter relation implies the inequality

$$\begin{aligned}
\|x^{i+n(i)} - x_i\| &\le 2r \, \gamma_i^{n(i)} + 2r \, (\gamma_{i+n(i)} - \gamma_i) \, (1 - \gamma_i)^{-1} \\
&= 2r \, \gamma_i^{n(i)} + (\varepsilon_i - \varepsilon_{i+n(i)}) \, \varepsilon_i^{-1},
\end{aligned}$$

in which the second term tends to zero for $i \to \infty$ by virtue of property 5) of admissible sequences.

Since by condition 6) from Definition 4.4,

$$\ln \gamma_i^{n(i)} = n(i) \, \ln \gamma_i = n(i) \cdot \varepsilon_i \, \frac{\ln(1 - \varepsilon_i)}{\varepsilon_i} \to -\infty$$

as $i \to \infty$, the first term also decreases to zero, i.e.,

$$\lim_{i\to\infty} \|x^{i+n(i)} - x_i\| = 0. \tag{4.6}$$

Using property 4) of admissible sequences, we conclude that there exists a unique integer j such that $j + n(j) \le k < j + 1 + n(j+1)$.

Now taking $m = j + n(j)$ and $i = j + 1$ in equality (4.5), we arrive at the following relation:

$$\begin{aligned}
&\|x^k - x_{j+1}\| \\
&\quad \le \gamma_{j+1}^{k-j-n(j)} \|x^{j+n(j)} - x_{j+1}\| + 2r\,(\gamma_k - \gamma_{j+1}) \cdot (1 - \gamma_{j+1})^{-1} \\
&\quad \le \|x^{j+n(j)} - x_j\| + \|x_j - \hat{x}\| + \|\hat{x} - x_{j+1}\| \\
&\qquad + 2r\,(\varepsilon_{j+1} - \varepsilon_{j+1+n(j+1)}) \cdot \varepsilon_{j+1}^{-1},
\end{aligned}$$

where $\hat{x}$ is the v_0-normal solution for the equation $x = T(x)$.

Joining (4.6), property 5) of admissible sequences, and Theorem 4.1, we conclude that $\|x^n - x_{j+1}\| \to 0$ as $n \to \infty$. Since $\|\hat{x} - x^k\| \le \|\hat{x} - x_{j+1}\| + \|x_{j+1} - x^k\|$, the limit is

$$\lim_{k\to\infty} \|\hat{x} - x^k\| = 0. \qquad \square$$

4.3 Asymptotic rule for stopping the iterations

The problem of the approximation of a fixed point of an operator T, i.e., the problem of solving the operator equation

$$x = T(x) \tag{4.7}$$

can be ill-posed. For example, for $T(x) = x - (Ax - y)$ or in more general form $T(x) = x - \beta\,(A^*Ax - A^*y)$, problem (4.7) is equivalent to solving the equation

$$Ax = y$$

that in the case of a discontinuous (unbounded) inverse operator A^{-1} is an ill-posed problem, i.e., its solution is unstable with respect to perturbations of the initial data of A, y. In this case, if the iterative process (4.1) with the perturbed data is applied, then for the process convergence it is necessary to formulate the *rule for stopping the iterations*, by relating the number of iterations with the level of accuracy.

Besides the exact scheme (4.1), consider its approximate implementation

$$\tilde{x}^{k+1} = \gamma_{k+1}\,\widetilde{T}(\tilde{x}^k) + (1-\gamma_{k+1})\,v_0 \tag{4.8}$$

with the same initial point x^0. Introduce the parameter δ that characterizes inaccuracy of the input data as

$$\|\widetilde{T}(\tilde{x}^k) - T(\tilde{x}^k)\| \le \varphi(\delta), \quad k = 0, 1, \dots \tag{4.9}$$

or

$$\|\widetilde{T}(x^k) - T(x^k)\| \le \varphi(\delta), \quad k = 0, 1, \dots, \tag{4.10}$$

where $\varphi(\delta) \to 0$ for $\delta \to 0$. We make sure that for some dependence of the number of iterations $k(\delta)$ on the inaccuracy δ, scheme (4.8) generates a *regularizing algorithm*, i.e., a strongly converging process as $\delta \to 0$ (Vasin, 1988).

Theorem 4.6. *Let the conditions of Theorem 4.5 and conditions* (4.9) *or* (4.10) *be satisfied, and in the latter case, let additionally, $\widetilde{T} \in \mathcal{K}$ be satisfied.*

Then for the relation $k(\delta)\cdot\varphi(\delta) \to 0$, $\delta \to 0$, of the parameters we have convergence: $\tilde{x}^{k(\delta)} \to \hat{x}$, where $\hat{x}$ is a point from $\mathrm{Fix}(T)$ *closest to v_0.*

Proof. We have the representation

$$\hat{x} - \tilde{x}^{k+1} = [\hat{x} - x^{k+1}] + [x^{k+1} - \tilde{x}^{k+1}],$$

where the first term in the right-hand side tends to zero by Theorem 4.5, and for the second term (with taking into account that $T \in \mathcal{K}$, $0 < \gamma_k \le 1$ and condition (4.9) is satisfied) the following estimate holds,

$$\begin{aligned}\|x^{k+1} - \tilde{x}^{k+1}\| &\le \gamma_{k+1}\,\|T(x^k) - T(\tilde{x}^k)\| + \gamma_{k+1}\,\|T(\tilde{x}^k) - \widetilde{T}(\tilde{x}^k)\| \\ &\le \|x^k - \tilde{x}^k\| + \varphi(\delta) \le (k+1)\,\varphi(\delta).\end{aligned}$$

Under condition (4.10) on the approximation and $\widetilde{T} \in \mathcal{K}$, we have a similar estimate for the second term,

$$\begin{aligned}\|x^{k+1} - \tilde{x}^{k+1}\| &\le \gamma_{k+1}\,\|T(x^k) - \widetilde{T}(x^k)\| + \gamma_{k+1}\,\|\widetilde{T}(x^k) - \widetilde{T}(\tilde{x}^k)\| \\ &\le \varphi(\delta) + \|x^k - \tilde{x}^k\| \le (k+1)\,\varphi(\delta).\end{aligned}$$

The theorem is proved. □

5 Fejér mappings and sequences

5.1 Definitions and general properties

Let $\mathcal{X}$ be a linear normed (real) space and $T : \mathcal{X} \to \mathcal{X}$ be a mapping. In Section 1, the definition of the Fejér mapping was given as a mapping for which the set $M = \mathrm{Fix}(T)$ of fixed points is not empty and the following relation holds,

$$\|T(x) - z\| < \|x - z\| \quad \forall z \in M \;\; \forall x \notin M.$$

For the set of all M-Fejér mappings the denoting $\mathcal{F}_M$ was introduced.

Definition 5.1. A sequence $x_k \subset \mathcal{X}$ is called *M-Fejér*, if

$$\|x_{k+1} - z\| < \|x_k - z\| \quad \forall z \in M \;\; \forall k = 0, 1, \ldots$$

(compare with (Motzkin and Schoenberg, 1954)).

Evidently, a M-Fejér sequence is bounded, i.e., $\sup_k \|x_k\| < \infty$. This property we shall use without further notice.

Lemma 5.2. *Let $\mathcal{X} = \mathcal{H}$ be a real Hilbert space. If $M \subset \mathcal{H}$ and $T : \mathcal{H} \to \mathcal{H}$, $T \in \mathcal{F}_M$, then the set M is convex and closed.*

Proof. Establish convexity of the set M. If it is not convex, then there exist $x, y \in M$ such that $z = (x + y)/2 \notin M$. Since z belongs to the segment $[x, y]$, the equality $\|x-z\|+\|z-y\| = \|x-y\|$ holds for any Hilbert space $\mathcal{H}$. By virtue of the Fejér property of the mapping T, we have

$$\|T(z) - x\| < \|z - x\|, \quad \|T(z) - y\| < \|z - y\|.$$

From the shown relations it follows

$$\|T(z) - x\| + \|T(z) - y\| < \|x - y\|.$$

On the other side, by the property of the norm we have

$$\|x - y\| \le \|T(z) - x\| + \|T(z) - y\|.$$

A contradiction is obtained, and this proves the convexity of the set M.

Further, let x' be a limit point of the set M. If to suppose that $x' \notin M$, i.e., $T(x') \ne x'$, $\|T(x') - x'\| = \varepsilon > 0$, then, taking the point $y \in M$ such that $2\|x' - y\| < \varepsilon$, we obtain

$$0 < \varepsilon = \|T(x') - x'\| \le \|T(x') - y\| + \|x' - y\| < 2\|x' - y\| < \varepsilon$$

which is impossible. Thus, the set M is closed. □

Lemma 5.3. *Let $M \subset \mathcal{H}$. If x_k is an M-Fejér sequence and x' and x'' are two different weak limits of some of its subsequences, then*

$$M \subset \{x : (x' - x'', x) = \alpha\}, \tag{5.1}$$

where α is some constant.

Proof. By virtue of the definition of the M-Fejér sequence we have

$$\lim_{k\to\infty} \|x_k - y\| = \lim_{k\to\infty} \|x_{j_k} - y\| = \lim_{k\to\infty} \|x_{i_k} - y\|,$$

where $x_{j_k} \rightharpoonup x'$, $x_{i_k} \rightharpoonup x''$, $y \in M$. From that it follows $(x_{j_k} - y, x_{j_k} - y) - (x_{i_k} - y, x_{i_k} - y) \to 0$ as $k \to \infty$, i.e.,

$$\|x_{j_k}\|^2 - \|x_{i_k}\|^2 - 2\,(x_{j_k}, y) + 2\,(x_{i_k}, y) \to 0, \quad k \to \infty.$$

Since $(x_{j_k}, y) \to (x', y)$ and $(x_{i_k}, y) \to (x'', y)$, then $\alpha_k = (1\,/\,2)(\|x_{j_k}\|^2 - \|x_{i_k}\|^2) \to (x' - x'', y)$. But $\lim_{k\to\infty} \alpha_k = \alpha$ does not depend on $y \in M$; therefore, $(x' - x'', y) = \alpha$ for all $y \in M$. □

Corollary 5.4. *If M contains interior points, then an M-Fejér sequence has a unique weak limit point.* □

Corollary 5.5. *The set M can not contain more than one weak limit point of a Fejér sequence.* □

Remark 5.6. If $M \subset \mathbb{R}^n$, then all limit points of an M-Fejér sequence x_k (in this case the notions of weak and strong limits coincide) lie on the sphere $S = \{x : \|x - y\| = R_y = \lim_{k\to\infty} \|x_k - y\|\}$, where $y \in M$. Consequently, if x' and x'' are two different limit points of the sequence under consideration, then M belongs to the locus of points that are on the equal distance from x' and x'', i.e.,

$$M \subset \left\{x : \left(x' - x'',\, x - \frac{x' + x''}{2}\right) = 0\right\}. \tag{5.2}$$

The mentioned properties are not obligatory for the case when $\mathcal{H}$ is an infinite-dimensional space. Actually, let $\mathcal{H} = l_2$, where l_2 be the space of numerical sequences $x = (\gamma_i)_{i=1}^{\infty} : \sum_{i=1}^{\infty} \gamma_i^2 < \infty$, $M = \{0\}$, where 0 is the zero element from l_2, and $e_i = (0, \dots, 0, \gamma_i = 1, 0, \dots, 0, \dots)$. Consider the sequence x_k defined by the relations

$$x_k = \begin{cases} \left(1 + \frac{1}{k}\right) e_k, & k = 2s, \\ \left(1 + \frac{1}{k}\right) e_1, & k = 2s + 1. \end{cases}$$

It is easy to see that x_k is M-Fejér, $x_{2s} \rightharpoonup 0$, and $x_{2s+1} \rightharpoonup e_1$. This shows that in the case of an infinite-dimensional space we have the following:

1) weak limit points of the sequence x_k do not obligatory lie on the sphere $S = \{x : \|x - y\| = R_y\}$;
2) the hyperplane corresponding to the equation $(x' - x'', x) = \alpha$ and containing the set M is not obligatory passes through the center of the segment $[x', x'']$;
3) the membership of one of the weak limit points of the sequence x_k to the set M does not imply its uniqueness.

Remark 5.7. If we speak only about strong limit points of an M-Fejér sequence x_k, then the properties formulated in Lemma 5.3 and its corollaries remain valid but with some enhances, namely: firstly, the hyperplane corresponding to the equation $(x' - x'', x) = \alpha$ from (5.1) will be given by equation (5.2). Secondly, if one x' of the strong limit points of the sequence x_k belongs to the set M, then such a point is unique, and we have $x_k \to x'$.

5.2 Examples of Fejér mappings

1. Let $\mathcal{H}$ be a Hilbert space and $M = \{x : x \in \mathcal{H}, (a, x) - b \leq 0\}$ be a half-space in $\mathcal{H}$, $a \neq 0$. Introduce the denoting $l(x) = (a, x) - b, l^+(x) = \max\{l(x), 0\}$. Define the mapping

$$P_M^\lambda(x) = x - \lambda \frac{l^+(x)}{\|a\|^2} a, \tag{5.3}$$

which, in its essence, is the metric projection of the point x onto the set M with the coefficient of relaxation $0 < \lambda < 2$ (see formula (3.7)). Note that for $\lambda = 1$, the mapping P_M^λ passes into the ordinary metric projection. In Lemma 3.13 it was established that for any arbitrary convex closed set M the mapping P_M^λ for $0 < \lambda < 2$ belongs to the class $\mathcal{P}_M^\nu$, therefore, this mapping is of Fejér type.

2. Let the following finite system of linear inequalities be given on a Hilbert space $\mathcal{H}$:

$$l_j(x) = (a_j, x) - b_j \leq 0, \quad j = 1, 2, \ldots, m, \tag{5.4}$$

where $a_j \neq 0$ for any j. Analogously to (5.3) suppose

$$T_j(x) = x - \lambda_j \frac{l_j^+(x)}{\|a_j\|^2} a_j, \quad \lambda_j \in (0, 2). \tag{5.5}$$

Consider the mappings

$$T(x) = \sum_{j=1}^{m} \alpha_j T_j(x), \quad \alpha_j > 0, \quad \sum_{j=1}^{m} \alpha_j = 1, \tag{5.6}$$

$$T(x) = T_m T_{m-1} \ldots T_1. \tag{5.7}$$

By Lemma 3.13 each of the mappings T_j is of M_j-Fejér type (moreover, $T_j \in \mathcal{P}_{M_j}^{\nu_j}$), where M_j is the half-space defined by the jth inequality in system (5.4). Let $M = \bigcap_{j=1}^m M_j \neq \varnothing$ be the polyhedron of solutions of system (5.4). Then each of the mappings (5.6), (5.7) is of M_j-Fejér type. This fact follows from the following general statement for arbitrary mappings $T_j \in \mathcal{F}_{M_j}$.

Theorem 5.8. *If $T_j \in \mathcal{F}_{M_j}$, $j = 1, 2, \ldots, m$, and $M = \bigcap_{j=1}^m M_j \neq \varnothing$, then each of the mappings T defined by formulas* (5.6) *and* (5.7) *belongs to the class $\mathcal{F}_M$ of Fejér mappings.*

Proof. This is checked directly by the scheme described in the proof of Theorem 1.8. □

Corollary 5.9. *If $\{T_j\}_1^n \in \mathcal{F}_M$, where M is an arbitrary convex closed subset of $\mathcal{H}$, then the* posynom

$$T(x) = \sum_{i=1}^{m} \alpha_i T_{i_1}^{n_{i_1}} T_{i_2}^{n_{i_2}} \ldots T_{i_m}^{n_{i_m}}(x)$$

is a Fejér mapping; here, $\alpha_i > 0$, $\sum_{i=1}^m \alpha_i = 1$.

Remark 5.10. The basic constructions of the M-Fejér mappings described in Theorem 5.8 give a broad range of possibilities for constructing new mappings from the class $\mathcal{F}_M$. The class $\mathcal{F}_M$ turns out to be closed with respect to the operations of superposition and convex combinations of any finite set of mappings from $\mathcal{F}_M$.

5.3 Weak-Fejér mappings and sequences

In Section 1 we introduced the class $\mathcal{K}_M$ of weakly M-Fejér (M-quasi-non-expansive) mappings T satisfying the conditions

$$\operatorname{Fix}(T) = M \neq \varnothing, \quad \|T(x) - z\| \leq \|x - z\| \quad \forall x \in \mathcal{X} \ \forall z \in M.$$

Definition 5.11. A linear normed space $\mathcal{X}$ is called *strongly convex* (normed), if the condition $\|x + v\| = \|x\| + \|v\|$, where $x \neq 0$, $v \neq 0$, implies that $v = cx, c > 0$.

Lemma 5.12. *Let $\mathcal{X}$ be a strongly convex space. If $T : \mathcal{X} \to \mathcal{X}$, $T \in \mathcal{K}_M$, then under $0 < \lambda < 1$ $T_\lambda = \lambda T + (1-\lambda) I \in \mathcal{F}_M$, i.e., this is an M-Fejér operator.*

Proof. Assume the contrary. Then for any $x \in \mathcal{X}$, $x \notin \operatorname{Fix}(T)$, $z \in \operatorname{Fix}(T) = M$ $\|T_\lambda(x) - z\| = \|x - z\|$ ($\|T_\lambda(x) - z\| > \|x - z\|$ is impossible by virtue of $T_\lambda \in \mathcal{K}_M$). We have

$$\begin{aligned}\|x - z\| = \|T_\lambda(x) - z\| &= \|\lambda\,(T(x) - z) + (1-\lambda)(x - z)\| \\ &\leq \|\lambda\,(T(x) - z)\| + \|(1-\lambda)(x - z)\| \leq \|x - z\|.\end{aligned}$$

From the strong convexity of $\mathcal{X}$, we conclude that for some $c > 0$

$$\begin{aligned}\lambda\,(T(x) - z) &= (1-\lambda)\,c\,(x - z), \\ \|T(x) - z\| &= \frac{c(1-\lambda)}{\lambda}\,\|x - z\|,\end{aligned}$$

i.e., $c(1-\lambda)\,/\,\lambda = 1$, $c = \lambda\,/\,(1-\lambda)$. Substituting the found value of c into the first equality of the two previous ones, we obtain the relation $T(x) = x$ that contradicts to the condition $x \notin \operatorname{Fix}(T)$.

Remark 5.13. The set of fixed points of a weakly M-Fejér mapping is convex and bounded (see Lemma 1.12).

Remark 5.14. If a nonexpansive mapping $T : \mathcal{X} \to \mathcal{X}$ (i.e., $\|T(x) - T(v)\| \leq \|x - v\|$) has at least one fixed point, i.e., $M = \operatorname{Fix}(T) \neq \varnothing$, then T is a weakly M-Fejér mapping; therefore, for a strongly convex space $\mathcal{X}$ we have $T_\lambda = \lambda T + (1-\lambda) I \in \mathcal{F}_M$.

Definition 5.15. A sequence x_k is called *weakly M-Fejér* if for all k we have $x_k \neq x_{k+1}$ and if in addition for any $y \in M$ the inequality $\|x_{k+1} - y\| \leq \|x_k - y\|$ holds (compare with (Motzkin and Schoenberg, 1954)).

For a weak M-Fejér sequence, Lemma 5.3, Corollaries 5.4 and 5.5, and Remark 5.6 formulated for the M-Fejér sequences are valid.

6 Theorems on convergence of Fejér processes

6.1 The case of the single-valued operator

To obtain the strong convergence of iterations in a Hilbert space, we need another additional property of the Fejér operator T.

Definition 6.1. Let an operator T act on the pair $\mathcal{X}, \mathcal{Y}$ of linear normed spaces. This operator T is called *completely continuous* if it transforms any sequence x_n weakly converging to $x_0 \in \mathcal{X}$ into the sequence $T(x_n)$ converging strongly to $T(x_0)$ in the space $\mathcal{Y}$.

Theorem 6.2. *Let $\mathcal{H}$ be a Hilbert space, $T : \mathcal{H} \to \mathcal{H}$, $T \in \mathcal{F}_M$, $M \subset \mathcal{H}$, and let the operator T be completely continuous. Then*

$$x^k = T^k(x^0) \to x' \in M, \tag{6.1}$$

where x^0 is an arbitrary element from $\mathcal{H}$.

Proof. Since we have $\|x^{k+1} - z\| \leq \|x^k - z\|$ for any $z \in M$, the limit $\lim_{k\to\infty} \|x^k - z\| = \rho(z) \geq 0$ exists. Since the sequence x^k is bounded and $\mathcal{H}$ is a Hilbert space, a weakly converging subsequence can be found,

$$x^{k_j} \rightharpoonup x'.$$

By virtue of the complete continuity of the operator T, the subsequence $x^{k_j+1} = T(x^{k_j})$ converges strongly,

$$x^{k_j+1} \to x'' = T(x').$$

Show now that $x' \in M$. Actually, if we assume that $x' \notin M$, then for $z \in M$ the chain of inequalities

$$\begin{aligned}\|x' - z\| &\leq \varliminf_{j\to\infty} \|x^{k_j} - z\| = \lim_{j\to\infty} \|T(x^{k_j}) - z\| \\ &= \|T(x') - z\| < \|x' - z\|\end{aligned}$$

leads to a contradiction. Thus, $T(x') = x'$ that implies $x^{k_j+1} \to x'$. If $\bar{k}$ is the largest number in the sequence $k_j + 1$ satisfying the condition $\bar{k} \leq k$, then the inequality $\|x^k - x'\| \leq \|x^{\bar{k}} - x'\|$ is valid, from which the strong convergence of the whole sequence $x^k \to x'$ as $k \to \infty$ follows. □

Corollary 6.3. *If $\mathcal{H}$ is finite-dimensional, then from the property of continuity of $T \in \mathcal{F}_M$, it follows* (6.1). □

Corollary 6.4. *Let x^k be an iterative sequence generated by a continuous M-Fejér mapping $T \in \mathcal{F}_M$, $M \subset \mathcal{H}$. If the linear hull of* $\mathrm{Lin}\{x^k\}_0^\infty$ *has finite dimension, then* (6.1) *is valid.* □

Corollary 6.5. *If a mapping $T \in \mathcal{F}_M$ is continuous, K is a convex compactum from $\mathcal{H}$, $\overline{M} = M \cap K \neq \varnothing$, and*

$$T_K(x) = P_K(T(x)), \tag{6.2}$$

where P_K is the metric projection onto the set K, then as $k \to \infty$

$$T_K^k(x_0) \to x' \in \overline{M}. \tag{6.3}$$

□

Remark 6.6. Instead of a convex compactum K in the previous corollary, it is also possible to use any convex closed *boundedly compact subset* of $\mathcal{H}$, i.e., a subset of the algebraic sum $Q + X_n$ of the absolutely convex compactum Q and the finite-dimensional subspace X_n (see Ivanov et al., 2002).

Example 6.7. Let M be a polyhedron given by a consistent system of linear inequalities (5.4), $T_j(x)$ be defined according to (5.5), and $T(x)$ be defined by (5.6). Since $\{T_j\}_1^k$ are continuous, then the operator T is also continuous. Further, from formulas (5.5), (5.6), it is seen that the elements of the sequence x^k generated by the iterative process $x^{k+1} = T(x^k)$ belongs to the linear hull of $\mathrm{Lin}\{x_0, a_j\}_{j=1}^n$. Therefore, the conditions of Corollary 6.4 are satisfied. That provides the convergence of (6.1).

6.2 The case of the multi-valued Fejér mappings

Definition 6.8. A *multi-valued* mapping $T : \mathcal{X} \to 2^{\mathcal{X}}$ is called *M-Fejér*, if

$$\forall\, z \in M,\ \ \forall\, x \notin M,\ \ \forall\, y \in T(x) \implies T(z) = z,\ \ \|y - z\| < \|x - z\|.$$

Let us reserve the denoting $\mathcal{F}_M$ for the class of the M-Fejér mappings both for one-valued and multi-valued ones. If some concrete mapping $T \in \mathcal{F}_M$ is multi-valued, then this fact will be noted specially.

In the case when a mapping T is multi-valued, then practically all results formulated above for the Fejér mappings remain valid with this or that change of the formulation.

Definition 6.9. A multi-valued mapping $T : \mathcal{X} \to 2^{\mathcal{X}}$ is called *closed*, if $x_k \in \mathcal{D}(T)$, $x_k \to x'$, $y_k \to y'$, $y_k \in T(x^k) \Longrightarrow x' \in \mathcal{D}(T)$, $y' \in T(x')$.

Note that if $\mathcal{X} = \mathbb{R}^n$ and T has the property

$$N \text{ is bounded} \Longrightarrow \bigcup_{x \in N} T(x) \text{ is bounded},$$

then in the case when T is single-valued and $\mathcal{D}(T) = \mathcal{X}$, the condition of closedness is equal to continuity.

For a multi-valued operator $T \in \mathcal{F}_M$, the M-Fejér sequences are generated according to the inclusion

$$x^{k+1} \in T(x^k). \tag{6.4}$$

Theorem 6.10. *Let $\mathcal{X} = \mathbb{R}^n$, $T : \mathbb{R}^n \to 2^{\mathbb{R}^n}$, $T \in \mathcal{F}_M$, and the mapping T be closed. Then the sequence x^k generated by relation* (6.4) *for an arbitrary x^0 converges to $x' \in M$.*

Proof. The theorem is proved by analogy to Theorem 6.2; for the sake of completeness the proof will be provided. Firstly, note that if $\bar{x} \in T(\bar{x})$, then $\bar{x} = T(\bar{x})$, i.e., $\bar{x} \in M$.

Actually, if $\bar{x} \notin M$, then by definition of a multi-valued M-Fejér mapping, we have for $y \in M$

$$\|z - y\| < \|\bar{x} - y\|$$

for all $z \in T(\bar{x})$. Taking $z = \bar{x}$, we obtain a contradictive inequality.

Now prove convergence of the sequence x^{k+1}, where $x^{k+1} \in T(x^k)$. From the boundedness of x^k, the existence of a converging subsequence x^{k_j}, $x^{k_j} \to x'$ follows. To avoid the additional separation of such a converging subsequence from x^{k_j+1}, let us suppose (without loss of generality) that $x^{k_j+1} \to x''$. Thus, we have the situation

$$\begin{array}{ccc} x^{k_j+1} & \in & T(x^{k_j}) \\ \downarrow & & \downarrow \\ x'' & & x' \end{array}$$

Then by virtue of the closedness of T, the inclusion $x'' \in T(x')$ holds. If $x' \in M$, then $x^k \to x'$ (see Corollary 5.5). But if $x' \notin M$, then we obtain $\|x'' - y\| < \|x' - y\|$ that contradicts to the idea of equal distance of the limit points of the Fejér sequence x^k from the point $y \in M$ (see Remark 5.6).

The theorem is proved. □

Example 6.11. For the system (5.4) of linear inequalities in $\mathbb{R}^n$ with the polyhedron $M \neq \varnothing$ of solutions, we consider the following mapping,

$$T(x) = \left\{ x - \lambda \frac{d^+(x)}{\|h\|^2} h : h \in \partial d(x) \right\}. \tag{6.5}$$

Here, $0 < \lambda < 2$, $d(x) = \max_{1 \le j \le m} l_j(x)$, $\partial d(x)$ is the subdifferential of the function $d(x)$; then, as it is known (Shor, 1979, Theorem 1.13), $\partial d(x) = \operatorname{conv}_{j \in J(x)} \{a_j\}$, $J(x) = \{j : d(x) = l_j(x)\}$.

Mapping (6.5) is closed (see Section 7 and Theorem 7.6). The generation of Fejér sequences with respect to the set M of solutions of system (5.4) by means of $T(x)$ can be obeyed to the mapping

$$T_0(x) = \left\{ x - \lambda \frac{d^+(x)}{\|a_j\|^2} a_j : j \in J(x) \right\} \tag{6.6}$$

with the narrower domain of values for each fixed x. The corresponding sequences $x^{k+1} \in T_0(x^k)$ will converge to elements from M according to Theorem 6.10.

7 M-separating pairs and M-Fejér mappings

Let M be a convex closed subset of $\mathbb{R}^n$, E be the mapping $\mathbb{R}^n \to 2^{\mathbb{R}^n}$, and $d(x)$ be a convex function with the property $\{x : d(x) \le 0\} = M$.

Definition 7.1. A pair $\{d(x), E(x)\}$ is called *M-separating* if for an arbitrary $p \notin M$ the half-space corresponding to the inequality

$$(h, x - p) + d(p) \le 0 \tag{7.1}$$

contains the set M for any $h \in E(p)$.

Example 7.2. Let $f(x)$ be a convex function and $M = \{x : f(x) \le 0\} \neq \varnothing$. Then $\{f(x), \partial f(x)\}$ is an M-separating pair; here, $\partial f(x)$ is the subdifferential of the function $f(x)$ at the point x, i.e.,

$$\partial f(x) = \{h : (h, y - x) \le f(y) - f(x)\}$$

for any $y \in \mathbb{R}^n$. The validity of relation (7.1) follows directly from the definition of a subdifferential.

Note properties of the subdifferential that are well known from convex analysis (see, for instance, (Polyak, 1983, Lemma 6, Chapter 5)): for any convex function $f : \mathbb{R}^n \to \mathbb{R}^1$ the subdifferential $\partial f(x)$ is a non-empty, convex, closed, and bounded set.

Now we give (in the form of lemmas) two additional important properties of the subdifferential as a mapping $\partial : x \to \partial f(x)$.

Lemma 7.3. *Let $f(x)$ be a convex function defined on $\mathbb{R}^n$ and*

$$\partial : x \to \partial f(x). \tag{7.2}$$

If N is a bounded subset of $\mathbb{R}^n$, then the set $\bigcup_{x \in N} \partial f(x)$ is bounded.

Proof. The proof is carried out by contradiction. Let there exist subsequences $x_k \subset N$ and h_k, $h_k \in \partial f(x_k)$, such that $\|h_k\| \to \infty$ for $k \to \infty$. By virtue of boundedness of N, the sequence x_k is bounded.

Suppose $h_k = t_k s_k$, $\|s_k\| = 1$. It is evident that $t_k \to \infty$. By the property of a subgradient of a convex function, we have

$$(h_k, x - x_k) \le f(x) - f(x_k)$$

for any x and k. Substituting $x = x_k + s_k$ into this inequality, we obtain

$$t_k \le \sup_k |f(x_k + s_k)| + \sup_k |f(x_k)| \le k < \infty.$$

Here, the fundamental property of a convex function is used, namely, its continuity at any point and, therefore, boundedness on any bounded closed set of $\mathbb{R}^n$. □

Lemma 7.4. *Let a function $f(x)$ defined on $\mathbb{R}^n$ be convex. Then the mapping (7.2) is closed.*

Proof. Actually, let $x_k \to x'$, $h_k \in \partial f(x_k)$, and $h_k \to h'$. It is necessary to show that $h' \in \partial f(x')$. Passing to the limit in the relation

$$(h_k, x - x_k) \le f(x) - f(x_k)$$

as $k \to \infty$, we obtain

$$(h', x - x') \le f(x) - f(x'),$$

that gives $h' \in \partial f(x')$. The lemma is proved. □

Let $\{d(x), E(x)\}$ be an M-separating pair. Construct the mapping (multi-valued in the general case)

$$T(x) = \left\{ x - \lambda \frac{d^+(x)}{\|h\|^2} h : h \in E(x) \right\}, \tag{7.3}$$

where $\lambda \in (0, 2)$. In (7.3) we suppose $T(x) = x$ if $d^+(x) = 0$, i.e., $d(x) \leq 0$. Note that if $x \notin M$, i.e., $d(x) > 0$, then for any h from $E(x)$ we have $h \neq 0$. Actually, if in this situation $h = 0$ would hold, then by property (7.1) we would have

$$(0, y - x) + d(x) \leq 0,$$

i.e., $d(x) \leq 0$ $(x \in M)$. But by the condition, $d(x) > 0$, then we obtain a contradiction. By this it is shown that (7.3) is defined correctly in the sense that zero can not appear in the denominator for $x \notin M$, but for $x \in M$, i.e., for $d(x) \leq 0$, we have $T(x) = x$ by definition. □

Lemma 7.5. *The mapping T given according to* (7.3) *is M-Fejér, i.e., $T \in \mathcal{F}_M$.*

Proof. This follows from the definition of an M-separating pair and Lemma 3.14. □

Theorem 7.6. *If in* (7.3) *the mapping $E(x)$ has the property of boundedness (i.e., it transforms a bounded set into a bounded one) and is closed, then $T(x)$ is closed.*

Proof. Let $x_k \to x'$, $y_k \to y'$, $y_k \in T(x^k)$. It is necessary to prove that $y' \in T(x')$. Consider two cases.

1) Let $x' \in M$, i.e., $d(x') \leq 0$. Since $T \in \mathcal{F}_M$, then $\|y_k - x'\| \leq \|x_k - x'\|$, consequently, $\|y_k - x'\| \to 0$ and $y' = x'$. But since $x' = T(x')$, the inclusion $y' \in T(x')$ holds automatically.

2) Let $x' \notin M$. Then for sufficiently large $k \geq \bar{k}$, the inequality $d(x_k) > 0$ is satisfied. Since $y_k \in T(x_k)$, we have

$$y_k = x_k - \lambda \frac{d(x_k)}{\|h_k\|^2} h_k, \quad h_k \in E(x_k), \tag{7.4}$$

and then, $h_k \neq 0$ (as it was noted above). Since the mapping $E(x)$ is bounded, the sequence h_k is bounded, and by this, it can be regarded converging, say to h'. Since E is closed, it holds $h' \in E(x')$. Passing to the limit as $k \to \infty$ in (7.4), we obtain

$$y' = x' - \lambda \frac{d(x')}{\|h'\|^2} h'; \tag{7.5}$$

i.e., $h' \in E(x')$, and relation (7.5) shows the inclusion $y' \in T(x')$, which was to be proved. □

Corollary 7.7. *Let $f(x)$ be a convex function defined on $\mathbb{R}^n$ and $M = \{x : f(x) \leq 0\} \neq \varnothing$. Then the mapping*

$$T(x) = \left\{ x - \lambda \frac{f^+(x)}{\|h\|^2} h : h \in \partial f(x) \right\} \tag{7.6}$$

is closed M-Fejér; by definition, for $f(x) \leq 0$, $T(x) = x$ for T as in (7.6) *or in* (7.3).

Proof. Actually, by virtue of Lemmas 7.3 and 7.4, all conditions of Theorem 7.6 hold for the situation of Corollary 7.7. □

Corollary 7.8. *Under the assumptions of Corollary 7.7 for the sequence x^k generated according to the inclusion $x^{k+1} \in T(x^k)$, the following convergence holds:*

$$x^k \to x' \in \{x : f(x) \leq 0\}. \tag{7.7}$$

Proof. Actually, since T is a closed mapping, convergence of (7.7) holds according to Subsection 6.10. □

Chapter II

Applications of iterative processes to nonlinear equations

1 Gradient methods

Consider a system of n nonlinear equations with n unknowns:

$$f_i(x) = 0, \quad i = 1, 2, \ldots, n, \tag{1.1}$$

where f_i are twice continuously differentiable functions. Below, for a short description of system (1.1) we shall use the operator form

$$F(x) = 0, \tag{1.2}$$

where $F : \mathbb{R}^n \to \mathbb{R}^n$. In the case of solvability, equation (1.2) is equivalent to the following minimization problem

$$\min \{\| F(x)\|^2 : x \in \mathbb{R}^n\}. \tag{1.3}$$

The necessary condition of an extremum for (1.3) takes the form

$$\nabla(\| F(x)\|^2) = 2J^\top(x)F(x) = 0,$$

where $J(x)$ denotes the Fréchet derivative of operator F, coinciding with the Jacobi matrix $J(x) = \{\partial f_i(x) / \partial x_j\}$ $(i, j = 1, 2, \ldots, m)$. By this, the *method of gradients* for solving problem (1.3) takes the form

$$x^{k+1} = x^k - \lambda J^\top(x^k)F(x^k). \tag{1.4}$$

To establish the convergence of iterations, let us ensure that under some conditions the step operator $T(x) = x - \lambda J^\top(x)F(x)$ belongs to the class $\mathcal{F}_M$ of Fejér mappings.

Theorem 1.1. *Let z be a solution of equation* (1.2) *and F be a twice continuously differentiable function that is vector-valued in the ball $S_r(z) = \{x \in \mathbb{R}^n : \|x - z\| \leq r\}$. Assume that the following conditions are satisfied:*

1) *for all $x \in S_r(z)$ there exists the inverse matrix J^{-1} and $\|J^{-1}(x)\| \leq B$;*

2) $\max_{k,j} \sum_{i=1}^{n} \left| \frac{\partial^2 f_k(x)}{\partial x_i \partial x_j} \right| \leq C$;

3) *the constants* r, B, C *satisfy the inequality* $n^{3/2} r B C < 2$.

Then the mapping

$$T(x) = x - \lambda J^\top(x) F(x)$$

for $0 < \lambda < 1 / \max\{\|J(x)\|^2 : x \in S_r(z)\}$ *belongs to the class* $\mathcal{F}_M$, *where* $M = \{z\}$, *i.e., this mapping is* M*-Fejér.*

Proof. Denote $R(x) = J(x)(x-z) - F(x)$. Taking into account that $F(z) = 0$, we have from the Taylor formula

$$\begin{aligned} \|R(x)\| &= \|F(x) - F(z) - J(x)(x-z)\| \\ &\leq \frac{1}{2} \max_\theta \|F''(\eta)(x-z, x-z)\|, \end{aligned}$$

where $\eta = z + \theta(x-z), 0 < \theta < 1$, $F''(\eta)$ is a bilinear mapping given by the formula

$$y_k = \sum_{i,j=1}^{n} a_{kij} x_i x_j, \quad a_{kij} = \frac{\partial^2 f_k(\eta)}{\partial x_i \partial x_j}.$$

Therefore, for $x \in S_r(z)$ we obtain the following estimate for $\|R(x)\|$:

$$\begin{aligned} \|R(x)\| &\leq \frac{1}{2} \max_\theta \Big[\sum_{k=1}^{n} \Big(\sum_{i,j}^{n} \frac{\partial^2 f_k(\eta)}{\partial x_i \partial x_j} (x_i - z_i, x_j - z_j) \Big)^2 \Big]^{\frac{1}{2}} \\ &\leq \frac{1}{2} \max_i |x_i - z_i|^2 \max_{0<\theta<1} \Big[\sum_{k=1}^{n} \Big(\sum_{i,j}^{n} \Big| \frac{\partial^2 f_k(\eta)}{\partial x_i \partial x_j} \Big| \Big)^2 \Big]^{\frac{1}{2}} \\ &\leq \frac{1}{2} \|x - z\|^2 \sqrt{n} \max_{0<\theta<1} \max_k \sum_{i,j=1}^{n} \Big| \frac{\partial^2 f_k(\eta)}{\partial x_i \partial x_j} \Big| \qquad (1.5) \\ &\leq \frac{1}{2} n^{3/2} \|x - z\|^2 \max_{0<\theta<1} \max_{k,j} \sum_{i=1}^{n} \Big| \frac{\partial^2 f_k(\eta)}{\partial x_i \partial x_j} \Big| \\ &\leq \frac{1}{2} n^{3/2} r^2 C. \end{aligned}$$

Taking into account the conditions of the theorem, we have

$$\begin{aligned}\|J(x)(x-z)\| &\geq (\|J^{-1}(x)\|)^{-1}\,\|x-z\| \\ &\geq B^{-1}\|x-z\| \\ &> \frac{1}{2}n^{3/2}\,r\,C\,\|x-z\| \\ &\geq \frac{1}{2}n^{3/2}\,\|x-z\|^2 C \geq \|R(x)\|\end{aligned} \tag{1.6}$$

for all $x \in S_r(z)$, $x \neq z$.

Since $F(x) = J(x)(x-z) + R(x)$, we obtain the estimate

$$\begin{aligned}&\|T(x)-z\|^2 \\ &\quad = \|x-z\|^2 - 2\lambda\,(J^\top(x)F(x), x-z) + \lambda^2\|J^\top(x)F(x)\|^2 \\ &\quad \leq \|x-z\|^2 - \lambda\,(\|F(x)\|^2 - \lambda\,\|J^\top(x)\|^2\,\|F(x)\|^2) < \|x-z\|^2.\end{aligned}$$

Here, the following inequality was used,

$$-2\lambda\,(J^\top(x)F(x), x-z) \leq -\lambda\,\|F(x)\|,$$

which follows from (1.6) and the condition of the theorem on the value of the parameter λ. □

Uniting Corollary 6.3 from Section 6 of Chapter I with Theorem 1.1, we obtain the following corollary.

Corollary 1.2. *Under the conditions of Theorem 1.1 for any initial approximation* $x^0 \in S_r(z)$, *the iterative process*

$$x^{k+1} = x^k - \lambda\,J^\top(x^k)F(x^k)$$

converges to the unique solution z *of equation* (1.2) *(of system* (1.1)*).*

Proof. Only the uniqueness of the solution in the ball $S_r(z)$ has to be revealed. Actually, the assumption on existence of another solution z' in the ball $S_r(z)$ implies $T(z') = z' - \lambda\,J'(z')F(z') = z'$, i.e., $\|T(z') - T(z)\| = \|z' - z\|$, which contradicts the theorem conclusion that $T \in \mathcal{F}_{\{z\}}$ in the ball $S_r(z)$. □

2 The Newton–Kantorovich method

This section is devoted to the investigation of this method to solve system (1.2). For this purpose we use the vector norm $\|x\| = \max_{1\le i\le n} |x_i|$ instead of the Euclidean norm. Then it is easy to check that the following estimate is valid for the residual term $R(x) = F(x) - F(z) - J(x)(x-z)$:

$$\|R(x)\| \le \frac{1}{2} n \, \|x-z\|^2 \max_{0<\theta<1} \max_{k,\,j} \sum_{i=1}^{n} \left| \frac{\partial^2 F_k(\eta)}{\partial x_i \partial x_j} \right|, \tag{2.1}$$

where $\eta = z + \theta(x-z)$.

Theorem 2.1. *Let z be a solution of the operator equation* (1.2) *and F be a vector-valued function twice continuously differentiable on the ball $S_r(z)$. Let the following conditions be also satisfied:*

1) *for all $x \in S_r(z)$ the inverse matrix $J^{-1}(x)$ exists and $\|J^{-1}(x)\| \le B$;*
2) $\max\limits_{k,i} \sum\limits_{j=1}^{n} |\partial^2 f_k(x) \,/\, \partial x_i \partial x_j| \le C$;
3) *the constants r, B, C satisfy the inequality $nrBC < 2$.*

Then the function $T(x) = x - J^{-1}(x)F(x)$ belongs to the class $\mathcal{F}_M$, where $M = \{z\}$.

Proof. From the conditions of the theorem and relation (2.1) the estimate follows:

$$\|R(x)\| \le \frac{1}{2} n \, C \, \|x-z\|^2.$$

Moreover, we have the relation

$$\|J^{-1}(x)R(x)\| \le \|J^{-1}(x)\| \, \|R(x)\| < \|x-z\|,$$

which implies

$$\begin{aligned} \|T(x) - z\| &= \|x - z - J^{-1}(x)F(x)\| \\ &= \|x - z - J^{-1}(x)\,[J(x)(x-z) + R(x)]\| \\ &= \|J^{-1}(x)R(x)\| \\ &< \|x-z\|, \end{aligned}$$

i.e., $T(x) = x - J^{-1}(x)F(x)$ belongs to the class $\mathcal{F}_{\{z\}}$ of Fejér mappings. □

Taking into account Theorem 6.2 (Corollary 6.3) from Chapter I, we obtain the corollary.

Corollary 2.2. *If the conditions of Theorem 2.1 are satisfied, equation* (1.2) *has a unique solution z in the ball $S_r(z)$, and the sequence x^k generated by the Newton–Kantorovich process*

$$x^{k+1} = x^k - J^{-1}(x^k)F(x^k), \quad x^0 \in S_r(z),$$

converges to z. □

The material of Sections 1 and 2 is based on the results from (Maruster, 1977).

3 Fejér processes for mixed problems

3.1 Systems of nonlinear equations and convex inequalities

In the two previous sections it was shown that the iterative operators in the gradient method and in the Newton–Kantorovich process are of Fejér type. This gives the opportunity for constructing Fejér processes for a more general problem than (1.1), namely:

Find a solution of the system

$$f_i(x) = 0, \ i = 1, 2, \dots, m, \quad f_i(x) \le 0, \ i = m+1, \dots, s, \tag{3.1}$$

i.e., a system of m nonlinear equations and $s - m$ inequalities is given.

Under the assumption of convexity and subdifferentiability of the functions f_i ($i = m+1, \dots, s$), let us associate the following mapping P with the mentioned system (3.1) of inequalities:

$$P(x) = \begin{cases} x - \lambda \dfrac{d(x)e(x)}{\|e(x)\|^2}, & d(x) > 0, \\ x, & d(x) \le 0. \end{cases} \tag{3.2}$$

Define the function $d(x)$ by one of the following formulas:

$$d(x) = \sum_{j=m+1}^{s} k_j \, [f_j^+(x)]^\mu, \quad k_j > 0, \ \mu > 1, \tag{3.3}$$

$$d(x) = \max_{m+1 \le j \le s} f_j(x), \tag{3.3 a}$$

and take some arbitrary subgradient of the function $d(x)$ at the point x as the function $e(x)$, i.e.,

$$e(x) \in \partial d(x). \tag{3.4}$$

It is clear that $Q = \{x : d(x) \leq 0\} = \mathrm{Fix}(P)$, where Q is the set of solutions of the system of inequalities. Let M be the set of solutions of the system of nonlinear equations in (3.1), and let $M \cap Q \neq \varnothing$.

Construct the iterative process

$$x^{k+1} = P(U(x^k)), \tag{3.5}$$

where U is the step operator in the method of gradients (i.e., $U(x) = x - \lambda J^{\top}(x)F(x)$) or in the Newton–Kantorovich method (i.e., $U(x) = x - J^{-1}(x)F(x)$), and the mapping P is defined by formulas (3.2), (3.3), and (3.4) or (3.2), (3.3 a), and (3.4).

Theorem 3.1. *Let the conditions of Theorems 1.1 or 2.1 be satisfied. Then the sequence x^k generated by process* (3.5) *converges to a solution of system* (3.1).

Proof. Actually, on the basis of Theorems 1.1 and 2.1, the operator U belongs to the class $\mathcal{F}_M$ and is continuous (since F is twice continuously differentiable). According to Lemma 7.4, Chapter I, the (multi-valued) mapping P is closed, and by virtue of Lemma 7.5, Chapter I, it belongs to the class $\mathcal{F}_Q$. Then $PU \in \mathcal{F}_{M\cap Q}$, and this mapping is closed as the superposition of a continuous operator and a closed operator. The reference to Theorem 6.10, Chapter I, completes the proof. □

Remark 3.2. Since under the conditions of Theorems 1.1 and 2.1 the solution of the system of equations

$$f_i(x) = 0, \quad i = 1, 2, \dots, m$$

is unique, the system of inequalities

$$f_i(x) \leq 0, \quad i = m + 1, \dots, s$$

can be interpreted as an additional *a priori* constraint on the solution of the system of equations.

An application of the process (3.5) instead of the gradient method or the Newton–Kantorovich can be more advisable, since by a small increase of computations (for the implementation of the operator P), an additional shift in direction to the solution is carried out in each step, and by this, the quality of the solution is improved without additional computational expenditures.

Remark 3.3. The mappings of the following form can be used as the operator P:

$$P(x) = \sum_{i=1}^{s} \lambda_i P_i(x), \quad P(x) = P_{j_1} P_{j_2} \dots P_{j_n}(x),$$

where P_i is a Q_i-Fejér mapping (Q_i is the set of solutions of the ith inequality in (3.1)) represented by the formula

$$P_i(x) = x - \lambda \frac{f_i^+(x)h}{\|h\|^2}, \quad 0 < \lambda < 2,$$

where $h \in \partial f_i(x)$.

3.2 Linear case

Now consider in problem (3.1) the special case that a system of m linear algebraic equations and $s - m$ linear inequalities is given, rewritten in the vector form

$$\begin{aligned} (x, a_i) - b_i &= 0, \quad i = 1, 2, \dots, m, \\ (x, a_i) - b_i &\leq 0, \quad i = m+1, \dots, s, \end{aligned} \tag{3.6}$$

where $x, a_i \in \mathbb{R}^n$, $b_i \in \mathbb{R}^1$. We suppose that $a_i \neq 0$, $i = 1, \dots, s$.

The metric projection P_{L_i} onto the hyperplane $L_i = \{x : (x, a_i) - b_i = 0\}$ ($i = 1, 2, \dots, m$) and onto the half-space $L_i = \{x : (x, a_i) - b_i \leq 0\}$ ($i = m+1, \dots, s$) is represented by the same formula,

$$P_{L_i} = \begin{cases} x - \dfrac{((x, a_i) - b_i)\, a_i}{\|a_i\|^2}, & x \notin L_i, \\ x, & x \in L_i. \end{cases} \tag{3.7}$$

Let $i_1, i_2, \dots, i_m$ be any ordering of the indices $1, 2, \dots, m$. Construct the iterative processes

$$x^{k+1} = P_{L_{i_1}} P_{L_{i_2}} \dots P_{L_{im}}(x^k), \quad x^0 \in \mathbb{R}^n, \tag{3.8}$$

$$x^{k+1} = \sum_{l=1}^{s} \lambda_l P_{L_{i_l}}(x^k), \quad \sum_{l=1}^{s} \lambda_l = 1, \ \lambda_l > 0, \ x^0 \in \mathbb{R}^n, \tag{3.9}$$

which converge to a solution of system (3.6) in the case of consistency (see Corollary 3.8 from Theorem 3.1, Chapter I). If the inequalities in problem (3.6) are missing, then the iterative processes (3.8) and (3.9) are known as the Kaczmarz algorithms (Kaczmarz, 1937).

4 Nonlinear processes for linear operator equations

4.1 Iterative α-processes and extremal principles

In Section 2, Chapter I, the iterative processes (of the types: the simple iteration method and the implicit scheme, i.e., the linear methods) have been already considered in a Hilbert space for the linear equation

$$Ax = y \tag{4.1}$$

with an operator A such that the inverse operator A^{-1} does not exist or is unbounded.

Now consider the class of nonlinear iterative methods (i.e., in which the step operator is nonlinear) for solving equation (4.1) with a bounded, selfadjoint, and semidefinite operator A. Such methods have the general name *α-processes* (see (Krasnosel'skii et al., 1969)).

Let α be some fixed real number from the interval $[-1, \infty)$. Define the iterative sequence by means of the recurrent formula

$$x^{k+1} = \begin{cases} x^k - \dfrac{(A^\alpha \Delta^k, \Delta^k)}{(A^{\alpha+1}\Delta^k, \Delta^k)}\Delta^k, & \Delta^k \neq 0, \\ x^k, & \Delta^k = 0, \end{cases} \tag{4.2}$$

where $\Delta^k = Ax^k - y$.

For $\alpha = 1$ we obtain the method of minimal residuals (Krasnosel'skii et al., 1969). For $\alpha = 0$, process (4.2) is transformed into the known method of steepest descent (see, for instance, (Kantorovich and Akilov, 1959)).

In the case $\alpha = -1$, (4.2) is the so called method of minimal errors. At first glance it seems that this method cannot be applied since in (4.2) the inverse operator A^{-1} appears. But, for an equation of the form $A^*Ax = A^*y$, this method can be transformed to the form, which is completely appropriate for realization (see below).

These three mentioned methods are constructed by the following common procedure. The following iterative scheme is taken as the initial one:

$$x^{k+1} = x^k - \beta_k(Ax^k - y),$$

and the parameter β_k is found by means of one of the following extremal principles.

If the parameter β_k is found from the extremal problem

$$\min_{\beta} \|Ax^{k+1} - y\|^2,$$

where $x^{k+1} = x^k - \beta\,(Ax^k - y)$, then we obtain the *method of minimal residuals*

$$x^{k+1} = x^k - \frac{(A(Ax^k - y),\ Ax^k - y)}{\|A(Ax^k - y)\|^2}\,(Ax^k - y).$$

If β_k is determined from the condition

$$\min_{\beta}\{(Ax^{k+1}, x^{k+1}) - 2\,(x^{k+1}, y)\},$$

then we obtain the *method of steepest descent*

$$x^{k+1} = x^k - \frac{\|Ax^k - y\|^2}{(A(Ax^k - y), Ax^k - y)}\,(Ax^k - y).$$

If β_k is determined from the condition

$$\min\ \|z - x^{k+1}\|^2,$$

where z is a solution of the equation $A^*Ax = A^*y$, $x^{k+1} = x^k - \beta\,(Ax^k - y)$, then we get the *method of minimal errors*,

$$x^{k+1} = x^k - \frac{\|Ax^k - y\|^2}{\|A^*(Ax - y)\|^2}\,A^*(Ax^k - y).$$

It is interesting to note that it is possible to obtain the method of minimal errors if we consider the basic Fejér mapping for the inequality

$$f(x) = \|Ax - y\|^2 \leq 0,$$

which is identical with equation (4.1) in the case of solvability. Actually, in this case the operator T defined by formula (3.11) (Section 3, Chapter I) takes the form

$$\begin{aligned} T(x) &= x - \lambda\,\frac{f(x)}{\|\nabla f(x)\|^2}\nabla f(x) \\ &= x - \frac{\lambda}{2}\,\frac{\|Ax - y\|^2}{\|A^*(Ax - y)\|^2}\,A^*(Ax - y). \end{aligned}$$

For $\lambda = 2$ (the limit value for the Fejér mapping T), we obtain the *method of minimal errors.*

Let us ensure that the step operator

$$T(x) = \begin{cases} x - \dfrac{(A^\alpha(Ax-y), Ax-y)}{(A^{\alpha+1}(Ax-y), Ax-y)}(Ax-y), & Ax \neq y, \\ x, & Ax = y \end{cases} \tag{4.3}$$

for the α-processes (4.2) is pseudo-contractive, i.e., it belongs to the class $\mathcal{P}^1_M$ and, consequently, is M-Fejér. Note that for an invertible operator A, for $Ax \neq y$ the denominator

$$(A^{\alpha+1}(Ax-y), Ax-y) \neq 0$$

and

$$\beta(x) = \frac{(A^\alpha(Ax-y), Ax-y)}{(A^{\alpha+1}(Ax-y), Ax-y)} \neq 0.$$

Thus, $\mathrm{Fix}(T) = \{\hat{x}\}$, where $\hat{x}$ is a solution of equation (4.1), and both the operator T and process (4.2) are correctly defined. In the case of a noninvertible operator A, it is necessary to require the condition $(A^{\alpha+1}(Ax-y), Ax-y) \neq 0$ for $Ax \neq y$ at least at the iteration points (or to consider a preliminary regularization, changing A to $A + \varepsilon I$).

4.2 Inequality for moments and pseudo-contractivity of the step operator

Below, we shall need a statement that touches one fact from the theory of moments for a selfadjoint positive semidefinite operator A.

Relations

$$b_s = (A^s x, x) = \int_m^M \lambda^s d\,(E_\lambda x, x),$$

where m, M are boundaries of the spectrum of the operator A, are called the *moments of an operator* A.

Let $s_1, s_2, \dots, s_k$ be arbitrary numbers and $\alpha_1, \dots, \alpha_k$ be positive ones. The pair of numbers $\{\omega, \tau\}$, where $\omega = \alpha_1 + \dots + \alpha_k$, $\tau = \alpha_1 s_1 + \dots + \alpha_k s_k$, is called the *dimension of the product* $b_{s_1}^{\alpha_1} b_{s_2}^{\alpha_2} \dots b_{s_k}^{\alpha_k}$.

Theorem 4.1. *The inequality*

$$b_{p_1}^{\alpha_1} b_{p_2}^{\alpha_2} \dots b_{p_k}^{\alpha_k} \leq b_{s_1}^{\beta_1} b_{s_2}^{\beta_2} \tag{4.4}$$

holds, if the dimensions of both sides are equal and if

$$s_1 < p_1, \dots, p_k < s_2.$$

Proof. A proof can be found in the book (Krasnosel'skii et al., 1969). □

Lemma 4.2. *Let M be a nonempty set of solutions for equation* (4.1), *and let A be a selfadjoint positive semidefinite operator acting in a Hilbert space $\mathcal{H}$. Then the operator T defined by formula* (4.3) *belongs to the class $\mathcal{P}^1_M$ (Definition 1.3, Chapter I).*

Proof. As noted in Section 1, Chapter I, relation (1.3) in the definition of the class $\mathcal{P}^1_M$ is equivalent to the inequality (1.4), Chapter I, which has the form

$$\|T(x) - z\|^2 \le (T(x) - z, x - z) \quad \forall\, z \in M, \quad \forall\, x \in \mathcal{H}. \tag{4.5}$$

Substituting relation (4.3) for the operator T into this inequality, we obtain the expression

$$\frac{(A^\alpha(Ax-y), Ax-y)}{(A^{\alpha+1}(Ax-y), Ax-y)} \left[\frac{(A^\alpha(Ax-y), Ax-y)}{(A^{\alpha+1}(Ax-y), Ax-y)} \times \|Ax-y\|^2 - (Ax-y, x-z) \right] \le 0. \tag{4.6}$$

Taking into account the notations introduced above for the operator moments and the fact that

$$\|Ax-y\|^2 = (A^0(Ax-y), Ax-y),$$
$$(Ax-y, x-z) = (A^{-1}(Ax-y), Ax-y),$$

relation (4.6) can be rewritten in the equivalent form

$$b_\alpha\, b_0 \le b_{-1}\, b_{\alpha+1}.$$

Validity of the latter inequality follows from Theorem 4.1 if we put $k = 2$, $s_1 = -1$, $s_2 = \alpha + 1$, $p_1 = \alpha$, $p_2 = 0$ in relation (4.4). By this, it is proved that relation (4.5) holds, i.e., $T \in \mathcal{P}^1_M$. □

4.3 Convergence of the α-processes

Now consider the condition of Theorem 3.9, Chapter I, on the weak convergence of the iterations for pseudo-contractive operators, to ensure that this theorem is valid for α-processes.

Lemma 4.3. *Let the operator A be invertible and for any j let $Ax_j - y \ne 0$. Then the operator T defined by relation* (4.3) *satisfies property* (3.1) *of Chapter I, i.e., $x_j \rightharpoonup \bar{x}$, $x_j - T(x_j) \to 0 \Rightarrow \bar{x} \in \mathrm{Fix}(T)$.*

Proof. Since for $Ax_j - y \neq 0$

$$
\begin{aligned}
|\beta_j| &= \frac{|(A^\alpha(Ax_j - y), Ax_j - y)|}{|(A^{\alpha+1}(Ax - y), Ax - y)|} \\
&= \frac{|(A^\alpha(Ax_j - y), Ax_j - y)|}{|A^{1/2}A^{\alpha/2}(Ax_j - y), A^{1/2}A^{\alpha/2}(Ax_j - y)|} \\
&\geq \frac{1}{\|A^{1/2}\|^2},
\end{aligned}
$$

then the relation $x_j - T(x_j) = \beta_j(Ax_j - y) \to 0$ implies $Ax_j - y \to 0$. By taking into account the property of weak continuity of the operator A, from $x_j \rightharpoonup \bar{x}$ we obtain $A\bar{x} = y$, i.e., $\bar{x} \in \mathrm{Fix}(T)$. □

Combining Lemmas 4.2, 4.3 with Theorem 3.1, Chapter I, we arrive at the following corollary.

Corollary 4.4. *Iterative α-processes* (4.2) *converge weakly to the solution $\hat{x}$ of equation* (4.1). □

Taking into account Theorem 3.9, Chapter I, we obtain the next corollary.

Corollary 4.5. *Let $\hat{x} \in Q$ (the a priori constraint), $P \in \mathcal{P}_Q^\nu$, and the operator T be defined by formula* (4.3). *Then the iterative method*

$$x^{k+1} = \widehat{T}(x^k),$$

where $\widehat{T} = PT$ or $\widehat{T} = \lambda P + (1-\lambda)T$, $0 < \lambda < 1$, converges weakly to the solution $\hat{x}$ of equation (4.1). □

Remark 4.6. If the operator A is positive definite, i.e., for some $c > 0$ we have $(Ax, x) \geq c\,\|x\|^2$, then α-processes converge strongly; see, for instance, (Krasnosel'skii et al., 1969).

To obtain a strong approximation of a solution in the general (incorrect) case, it is possible to proceed in the following way. Regularize equation (4.1) by the Lavrent'ev scheme (Lavrent'ev, 1962)

$$Ax + \varepsilon x = y. \tag{4.7}$$

As it is known, for a selfadjoint, positive semidefinite operator A and $\varepsilon > 0$, the solution x^ε of equation (4.7) exists, is unique, and for $\varepsilon \to 0$ converges to the normal solution of equation (4.1) both for exact and approximate data;

see (Lavrent'ev, 1962; Ivanov et al., 2002). Since the operator $A_\varepsilon = A + \varepsilon I$ is positive definite ($c = \varepsilon$), the α-processes with the operator A_ε converge strongly to the solution x^ε of equation (4.7). Thus, we obtain a two-stage regularizing algorithm for solving the original equation (4.1).

Note that modifying the α-processes on the basis of the principle of the iterative regularization, it is possible to obtain a one-stage iterative method that converges strongly; see (Bakushinskii and Goncharskii, 1994).

5 Linearized versions of the gradient methods

5.1 The method of steepest descent

Consider the nonlinear equation

$$A(x) = y \tag{5.1}$$

with the Fréchet differentiable operator A acting on the pair of Hilbert spaces $\mathcal{H}_1$ and $\mathcal{H}_2$, in the particular (finite-dimensional) case, problem (5.1) can be represented in the form of a system of nonlinear equations (3.1). If problem (5.1) is solvable, then this is equivalent to the problem of minimization

$$\min \left\{ \frac{1}{2} \|A(x) - y\|^2 : x \in \mathcal{H}_1 \right\}.$$

The necessary condition fo an extremum for this problem leads to the operator equation

$$A'(x)^*(A(x) - y) = 0, \tag{5.2}$$

which contains the set of solutions of equation (5.1), but can have additional solutions. Take the notations M and M^* for the set of solutions of equations (5.1) and (5.2), correspondingly.

Iterative processes of the form

$$x^{k+1} = x^k - \beta_k A'(x^k)^*(A(x^k) - y) \equiv T(x^k) \tag{5.3}$$

have to be regarded as gradient methods, since

$$S(x) \equiv A'(x)^*(A(x) - y) = \nabla \left[\frac{1}{2} \|A(x) - y\|^2 \right].$$

In the classical method of steepest descent, the parameter β_k is found from the extremal problem

$$\beta_k: \qquad \min_\beta \|A(x^k - \beta\, S(x^k)) - y\|^2,$$

i.e., at each step of the implementation of the method, it is necessary to solve a one-dimensional optimization problem. To avoid this additional procedure, let us modify the process in the following way. Linearize the operator A at the current point x^k

$$A(x) \approx A(x^k) + A'(x^k)(x - x^k)$$

and find the parameter β_k by minimization the square residual for the linearized equation,

$$\beta_k: \qquad \min_{\beta} \|A'(x^k)(x^k - \beta\, S(x^k)) - (y + A'(x^k)x^k - A(x^k))\|^2,$$

and this gives $\beta(x_k) = \|S(x^k)\|^2 \,/\, \|A'(x^k)\, S(x^k)\|^2$.

Thus, process (5.3) takes the form

$$x^{k+1} = \begin{cases} x^k - \dfrac{\|S(x^k)\|^2}{\|A'(x^k)S(x^k)\|^2} S(x^k) \equiv T(x^k), & S(x^k) \neq 0, \\ x^k, & S(x^k) = 0. \end{cases} \tag{5.4}$$

By analogy with the linear case, process (5.4) is called the *method of steepest descent* (MSD) (see, for instance, (Neubauer and Scherzer, 1995)).

Introduce the additional (local) condition on the operator A. Assume that for some constant $\kappa > 0$ in the neighborhood $S_\rho(z)$ of a solution $z \in M$ the following inequality holds,

$$\|A(x) - A(z)\|^2 \leq \kappa(A(x) - A(z),\, A'(x)(x - z)), \tag{5.5}$$

and, also, the condition $S(x) \notin \ker A'(x)$ is satisfied for $S(x) \neq 0$. Note that inequality (5.5) implies, particularly, coincidence of the solution sets $M = M^*$, and the condition $S(x) \notin \ker A'(x)$ for $S(x) \neq 0$ excludes that the denominator in process (5.4) vanishes.

For expanding the domain of convergence of the iterative method (5.4), introduce the additional control parameter γ and consider the process (Vasin, 1998)

$$x^{k+1} = x^k - \gamma\, \beta(x_k)\, S(x^k) \equiv T(x^k). \tag{5.6}$$

Lemma 5.1. *Let condition* (5.5) *hold. Then for* $\gamma < \frac{2}{\kappa}$ *the step operator* T *of process* (5.4) *belongs to the class* $\mathcal{P}^{\nu}_{M \cap S_\rho(z)}$, *where* $\nu = 2\,/\,\gamma\kappa - 1$.

Proof. This follows from the chain of relations

$$\begin{aligned}
&\|T(x)-z\|^2-\|x-z\|^2+\nu\,\|T(x)-x\|^2\\
&\quad=-2\gamma\,\beta(x)(S(x),x-z)+(1+\nu)\,\gamma^2\,\beta^2(x)\,\|S(x)\|^2\\
&\quad=(1+\nu)\,\gamma^2\,\beta(x)\Big[-\frac{2}{\gamma\,(1+\nu)}\,(S(x),x-z)+\frac{\|S(x)\|^4}{\|A'(x)\,S(x)\|^2}\Big]\\
&\quad\le(1+\nu)\,\gamma^2\,\beta(x)\Big[-\frac{2}{\gamma\,(1+\nu)}\,(S(x),x-z)+\|A(x)-y\|^2\Big]\\
&\quad\le(1+\nu)\,\gamma^2\,\beta(x)\big[-\kappa\,(S(x),x-z)+\|A(x)-y\|^2\big]\le 0.
\end{aligned}$$

□

5.2 Linearized analogue of the minimal error method

As in the previous section, linearize the nonlinear operator A at the point x^k. Denote the solution of the obtained linear equation

$$A'(x^k)x=y+A'(x^k)x^k-Ax^k$$

by $\bar{z}$. Define the parameter β_k in the iterative process (5.3) from the condition

$$\beta_k:\qquad \min_{\beta}\,\|x^{k+1}-\bar{z}\|^2,$$

this gives $\beta_k=\|A(x^k)-y\|^2\,/\,\|S(x^k)\|^2$. We obtain the iterative process

$$x^{k+1}=x^k-\frac{\|A(x^k)-y\|^2}{\|S(x^k)\|^2}\,A'(x^k)^*(A(x^k)-y),\tag{5.7}$$

which is called the *method of the minimal error* for a nonlinear operator equation. Introduce the additional parameter γ, which regulates the step value, and, finally, consider the process (Vasin, 1998)

$$x^{k+1}=x^k-\gamma\,\frac{\|A(x^k)-y\|^2}{\|S(x^k)\|^2}\,S(x^k)\equiv T(x^k).\tag{5.8}$$

Lemma 5.2. *If condition* (5.5) *is satisfied, then for* $\gamma<\frac{2}{\kappa}$ *the step operator* T *for process* (5.8) *belongs to the class* $\mathcal{P}^{\nu}_{M\cap S_\rho(z)}$ *of pseudo-contractive mappings, where* $\nu=2\,/\,\gamma\kappa-1$.

Proof. This is implemented by the scheme of the proof of Lemma 5.1. □

5.3 Conclusions and applications

The properties of the iterative operators established in Lemmas 5.1 and 5.2 in the methods of the steepest descent and the minimal errors allows one:

1) on the basis of Theorem 3.1, Chapter I, to make a conclusion on the weak convergence of these processes;
2) having an *a priori* information, to construct superpositions (or convex combinations) of the projection operators or operators of type (3.2) that are responsible for the *a priori* constraints;
3) to generate, together with other processes having the property of pseudo-contractivity, new classes of iterative methods for the approximative solution of equation (5.1).

Consider some examples of operators A satisfying property (5.5). First of all, note that for the linear operator A this property is evidently satisfied.

Example 5.3. Let for the operator A the residual $f(x) = \|A(x) - y\|^2$ be a convex functional. Then condition (5.5) will be satisfied for $\kappa = 2$, since this is the criterion of convexity for the differentiable functional $f(x)$. Particularly, if in system (1.1) the functions f_i are convex and nonnegative, then the residual $f(x) = \|A(x)\|^2$ for the operator $A(x) = (f_1(x), f_2(x), \dots, f_n(x))^\top$ is a convex function, hence, condition (5.5) holds.

Example 5.4. The inverse (coefficient) problem for the differential equation

$$-(a\,u_s)_s = f, \quad s \in (0,1) \tag{5.9}$$

with boundary conditions

$$u(0) = g_0, \quad u(1) = g_1 \tag{5.10}$$

can be formulated in the following way:

It is necessary to find the diffusion coefficient $a(s)$ by the function $u(s)$ known on $[0,1]$. Problem (5.9), (5.10) can be formally written as a nonlinear operator equation

$$A(a) = u(a),$$

where $u(a)$ is a solution of the boundary problem (5.9), (5.10). It is supposed that $f \in L_2$ and the operator A acts from $\mathcal{D}(A) = \{a \in \mathcal{H}^1[0,1],\ a(s) \ge \underline{a} > 0\}$ into $L_2[0,1]$.

It turns out that under these conditions the operator A satisfies the inequality

$$\|A(x) - A(\tilde{x}) - A'(x)(x - \tilde{x})\| \le \eta\,\|A(x) - A(\tilde{x})\| \tag{5.11}$$

for $\eta < 1/2$, and for all x, $\tilde{x}$ from some ball $S_\rho(a_0)$; details for study of this property can be found in (Neubauer and Scherzer, 1995; Engl et al., 1996).

It is easy to check that for $\tilde{x} = z$ inequality (5.11) implies property (5.5) with $\kappa = 2/(1-\eta^2)$; therefore, the operator A is pseudo-contractive.

Example 5.5. Consider the nonlinear Volterra equation

$$A(x)(t) \equiv \int_0^t \phi(x(s))\,ds = y(t) \tag{5.12}$$

under the assumption that $\phi \in C^2(\mathbb{R})$, $A : \mathcal{H}^2[0,1] \to L_2[0,1]$. It is directly verified that

$$(A'(x)\,h)(t) = \int_0^1 \phi'(x(s))\,h(s)\,ds,$$

$$((A'(x))^* u)(s) = B^{-1}\left[\phi'(x(s)) \int_s^1 u(t)\,dt\right],$$

where $B : \mathcal{D}(B) = \{\psi : \psi \in \mathcal{H}^1[0,1],\ \psi'(0) = \psi'(1) = 0\} \to L_2[0,1]$, $B\psi = -\psi'' + \psi$.

If $\phi'(x) \geq \kappa > 0$ for all x from the ball $S_\rho(x_0)$, then we have the representation

$$A'(v) = R_v(x)A'(x) \quad \forall\, x,\ v \in S_\rho(x_0),$$

where

$$R_v(x)^* w = -\left(\frac{\phi'(v)}{\phi'(x)} \int_s^1 w(t)\,dt\right)',$$

and the following estimate holds,

$$\|R_v(x) - I\| \leq C\,\|v - x\| \quad \forall x, v \in S_\rho(x_0). \tag{5.13}$$

Here, C is a positive constant that does not depend on v, x.

As was established in the article (Hanke et al., 1995), property (5.13) implies inequality (5.11), and, hence, the validity of inequality (5.5) guarantees the pseudo-contractive property of the step operator T in the processes (5.6) and (5.8).

Note that if inequality (5.11) holds, then the strong convergence of the steepest descent method and the minimal errors method is valid (see (Neubauer and Scherzer, 1995)).

6 The Levenberg–Marquardt method

6.1 The idea of the method

For approximately solving the nonlinear equation

$$A(x) = y, \tag{6.1}$$

given on a pair of Hilbert spaces $\mathcal{H}_1, \mathcal{H}_2$, consider an iterative process in the form

$$x^{k+1} = x^k + \triangle^k. \tag{6.2}$$

Linearize the operator A at the current point x^k

$$A(x) \simeq A(x^k) + A'(x^k)(x - x^k)$$

and find the unknown correction $\triangle^k$ from the assumption that each next approximation x^{k+1} minimizes the regularized residual, i.e.,

$$\min_{\triangle^k} \|A'(x^k)\, \triangle^k - [y - A(x^k)]\,\|^2 + \alpha\, \|x^{k+1} - x^k\|^2.$$

From this, we obtain the following expression for $\triangle^k$,

$$\triangle^k = -[A'(x^k)^* A'(x^k) + \alpha\, I\,]^{-1} A'(x^k)^* (A(x^k) - y)$$

and, therefore, process (6.2) takes the form

$$x^{k+1} = x^k - [A'(x^k)^* A'(x^k) + \alpha\, I\,]^{-1}\, A(x^k)^* (A(x^k) - y); \tag{6.3}$$

in the scientific literature, this process is sometimes called the *Levenberg–Marquardt method.*

Note that for $\alpha = 0$, the Levenberg–Marquardt process becomes the classical Gauss–Newton method; thus, it is also reasonable to call this process the *regularized analogue of the Gauss–Newton method.*

6.2 Weak convergence of the method

Before going to study convergence of the iterations, modify (6.3) somewhat by introduction of the parameter β that governs the step size, and assuming the parameter α to be variable. Thus, process (6.3) takes the form (Vasin and Mokrushin, 2000)

$$x^{k+1} = x^k - \beta\, [A'(x^k)^* A'(x^k) + \alpha_k I\,]^{-1} \times A(x^k)^* (A(x^k) - y). \tag{6.4}$$

As in the gradient methods, the following condition (see (5.5)) for the operator A is considered,

$$\|A(x) - y\|^2 \le \kappa\,(x - z, S(x)) \tag{6.5}$$

in some neighborhood $S_r(z)$ ($r = \|z - x^0\|$) of the solution z of equation (6.1); here, $S(x) = A(x)^*(A(x) - y)$.

Introduce the variable Hilbert norm

$$\|x\|_k^2 = (B_k x, x), \quad B_k = A'(x^k)^* A'(x^k) + \alpha_k I. \tag{6.6}$$

If we suppose that $N_1 = \sup_x \|A'(x)\| < \infty$, $\alpha_k \ge \alpha > 0$, then from the evident estimate

$$\alpha\,\|x\|^2 \le \|x\|_k^2 \le (N_1^2 + \alpha)\,\|x\|^2$$

it follows that for any k the introduced norm $\|x\|_k$ is equivalent to the original Hilbert norm of the space $\mathcal{H}_1$.

In the sequel, we also assume that in the ball $S_r(z)$ there are no other solutions of equation (6.1) except z.

Theorem 6.1. *Let, together with the assumptions mentioned above in this section, the relation*

$$x_k \rightharpoonup \bar{x}, \quad S(x_k) \equiv A'(x_k)^*(A(x^k) - y) \to 0 \Rightarrow S(\bar{x}) = 0$$

be satisfied. Then for $0 < \beta < 2\alpha / \kappa N_1^2$ *the iterations* x^k *generated by process* (6.4) *have the following properties:*

1) $x^k \rightharpoonup z$ *in the space* $\mathcal{H}_1$*, where* z *is a solution of equation* (6.1)*;*
2) *either* $\|x^{k+1} - z\|_k < \|x^k - z\|_k$ *for any* k*, or the sequence* x^k *is stationary under* $k \ge k_0$.

Proof. Introduce the notations $T_k(x) = x - \beta B_k^{-1} S(x)$, and let M be the set of solutions of equation (6.1). Note that with the given notations, we have $\mathrm{Fix}(T_k) = M$ for any k. Actually, for $\beta > 0$ the relation $T_k(x) = x$ implies $B_k^{-1} S(x) = 0$, from which $S(x) = 0$, but then $Ax - y = 0$ by virtue of (6.5), i.e., $x \in M$.

Now show that for $\nu > 0$ and for any k, the inequality holds

$$\|T_k x^k - z\|_k^2 - \|x^k - z\|_k^2 + \nu\,\|T_k(x^k) - x^k\|_k^2 \le 0. \tag{6.7}$$

After substitution of the expression for the operator T_k and evident transformations, inequality (6.7) takes the form

$$\left[-\frac{2}{(1+\nu)\,\beta}\,(x - z, S(x^k)) + (S(x^k), B_k^{-1} S(x^k))\right] \le 0.$$

If we require the relation

$$\left[-\frac{2}{(1+\nu)\beta}(x-z,S(x))+\frac{N_1^2}{\alpha}\,\|Ax-y\|^2\right]\le 0 \tag{6.8}$$

for all $x \in S_r(z)$, then this implies the previous inequality.

In the turn, taking into account property (6.5), inequality (6.8) will be satisfied if

$$\frac{2\alpha}{(1+\nu)\,\beta\,N_1^2}\ge\kappa.$$

So it is possible to take $\nu=\frac{2\alpha}{\beta N_1^2\kappa}-1$.

Thus, inequality (6.7) is established. From this it follows that there exists

$$\lim_{k\to\infty}\|x^k-z\|_k=d(z)\ge 0.$$

Then the sequence $\{\|x^k\|_k\}$ is bounded, and from this, by virtue of the norm equivalence, we obtain boundedness of $\{\,\|x^k\|\,\}$ in the space $\mathcal{H}_1$; therefore, for some subsequence

$$x^{k_i}\rightharpoonup\bar{x}. \tag{6.9}$$

Moreover, from (6.7) it follows that

$$\|T_k(x^k)-x^k\|_k=\beta\,\|B_k^{-1}S(x^k)\|_k\to 0,$$

and from this, taking into account the conditions of the theorem, we have $S(x^k)\to 0$ in the space $\mathcal{H}_1$. Gathering the latter relation with (6.9), we conclude that $S(\bar{x})=0$. Since the solution of equation (6.1) is unique, it implies $\bar{x}=z$ and the weak convergence is valid for the whole sequence,

$$x^k\rightharpoonup z.$$

By this, property 1) is proved.

Property 2) also holds, since if $T_kx^k\ne x^k$, it follows from relation (6.7)

$$\|x^{k+1}-z\|_k<\|x^k-z\|_k.$$

In the contrary case, if $T_{k_0}x^{k_0}=x^{k_0}$, then it means that $x^{k_0}=z$, therefore, for any $k\ge k_0$, $T_k(z)=z-B_k^{-1}S(z)=z$, i.e., the sequence x^k becomes stationary. □

6.3 Strong convergence of the modified method. Asymptotic rule for stopping the process

According to the considered notations, method (6.3) can be written in the form

$$x^{k+1} = x^k - \beta\, B_k^{-1} S(x^k) \equiv T_k(x^k). \tag{6.10}$$

If some additional information is known for equation (6.1), i.e., the solution z to be found belongs to a compact or boundedly compact set Q, then it is possible to construct a superposition of the operator T_k with the metric projection P_{Q_k} and to consider the process

$$x^{k+1} = (P_Q)_k\, T_k(x^k) \tag{6.11}$$

for an approximate given right-hand side of equation (6.1): $\|y - y_\delta\| \le \delta$.

In this case, process (6.11) is written in the form

$$\tilde{x}^{k+1} = (P_Q)_k\, \widetilde{T}_k(\tilde{x}^k). \tag{6.12}$$

Here, $(P_Q)_k$ is the projection operator in the space with the variable norm that was introduced above.

Theorem 6.2. *Let the conditions of Theorem 6.1 be satisfied and* $\|A'(x)\| \le N_1$, $\|A''(x)\| \le N_2$ *for* $x \in Q$, *where* Q *be a boundedly compact set. Let*

$$C = [(N_1^2 + \alpha)\,\alpha]^{1/2}\,[1 + 2\,N_1^3 N_2\, a \,/\, \alpha^2 + (N_1^2 + N_1 N_2\, a)\,/\,\alpha\,],$$

where $\|x^k - z\| \le a$.

Then if the number of iterations $k(\delta)$ *is chosen such that* $C^{k(\delta)}\,\delta \to 0$ *for* $\delta \to 0$, *then*

$$\lim_{\delta\to 0} \|\tilde{x}^{k(\delta)} - z\| = 0,$$

i.e., process (6.12) *generates a regularizing algorithm.*

Proof. Since the superposition of pseudo-contractive mappings will belong to the same class as process (6.10), method (6.11) converges weakly to z. And since the iterative sequence is bounded and belongs to the boundedly-compact set Q, this implies strong convergence of x^k to z.

From the triangle inequality for norms, we have

$$\begin{aligned}
&\|z - \tilde{x}^{k+1}\|_k \\
&\quad \le \|z - x^{k+1}\|_k + \|x^{k+1} - \tilde{x}^{k+1}\|_k \\
&\quad \le \|z - x^{k+1}\|_k \\
&\qquad + (\|T_k(x^k) - T_k(\tilde{x}^k)\| + \|T_k(\tilde{x}^k) - \widetilde{T}_k(\tilde{x}^k)\|)\,(N_1^2 + \alpha)^{1/2}.
\end{aligned}$$

By virtue of the fact established above, the first term at the right-hand side tends to zero. Estimate the second term:

$$
\begin{aligned}
&\|T_k(x^k) - T_k(\tilde{x}^k)\| \\
&\qquad = \|(x^k - \tilde{x}^k) - [B_k^{-1} S(x^k) - \widetilde{B}_k^{-1} S(\tilde{x}^k)\| \\
&\qquad \le \|x^k - \tilde{x}^k\| + \|B_k^{-1}[S(x^k) - S(\tilde{x}^k)]\| + \|(\widetilde{B}_k^{-1} - B_k^{-1}) S(x^k)\|.
\end{aligned}
$$

Here,

$$
\begin{aligned}
\|B_k^{-1}\| &\le 1/\alpha; \\
\|S(x^k) - S(\tilde{x}^k)\| &= \|A'(x^k)^* (A(x^k) - y) - A'(\tilde{x}^k)^* (A'(\tilde{x}^k) - y)\| \\
&\le \|A'(x^k)^* (A(x^k) - y) - A'(\tilde{x}^k)^* (A(x^k) - y)\| \\
&\qquad + \|A'(\tilde{x}^k)^* (A(\tilde{x}^k) - A(x^k))\| \\
&\le N_1 N_2\, a\, \|x^k - \tilde{x}^k\| + N_1^2 \|x^k - \tilde{x}^k\|; \\
\|x^k - z\| &\le a; \\
\|\widetilde{B}_k^{-1} - B_k^{-1}\| &= \|\widetilde{B}_k^{-1} (B_k - \widetilde{B}_k) B^{-1}\| \le \frac{2 N_1 N_2}{\alpha^2} \|\tilde{x}^k - x^k\|; \\
\|S(x^k)\| &= \|A'(x^k)^* (A(x^k) - y)\| \le N_1^2\, a.
\end{aligned}
$$

Now consider the third term. We have the evident estimate

$$
\begin{aligned}
&\|T_k(\tilde{x}^k) - \widetilde{T}_k(\tilde{x}^k)\| \\
&\qquad \le \|\widetilde{B}_k^{-1}\| \, \|A'(\tilde{x}^k)^* (A'(\tilde{x}^k) - y) - A'(\tilde{x}^k)^* (A(\tilde{x}^k) - y_\delta)\| \le \frac{N_1 \delta}{\alpha}.
\end{aligned}
$$

Gathering the obtained relations, we obtain the final estimate

$$
\|x^{k+1} - \tilde{x}^{k+1}\|_k \le \|x^k - \tilde{x}^k\|_k \, C + \psi(\delta) \le \Big(\sum_{i=0}^{k+1} C^i\Big)\, \psi(\delta),
$$

where $C = [(N_1^2 + \alpha)\,\alpha]^{1/2}\,[1 + 2 N_1^3 N_2\, a / \alpha^2 + (N_1^2 + N_1 N_2\, a)/\alpha]$ and $\psi(\delta) = (N_1^2 + \alpha)^{1/2} N_1 \delta / \alpha$; from this estimate, the necessary relation follows. □

6.4 Stopping the iterations by the residual

Let Q be a compact set. Let, for the given level of the error δ, the number of iterations $k(\delta)$ in process (6.11) be defined from the relations

$$\|A(\tilde{x}^{k+1}) - y_\delta\| \le \tau\,\delta < \|A(\tilde{x}^{k}) - y_\delta\|,$$
$$k = 1, 2, \ldots, k(\delta) - 1, \quad \tau > 1.$$

Under the assumption of unique solvability (i.e., the operator A is invertible) of equation (6.1) such a number $k(\delta)$ exists for any δ and $k(\delta) \to \infty$ for $\delta \to 0$, then the sequence $x^{k(\delta)}$ converges to the solution z, i.e.,

$$\lim_{\delta \to 0} \|\tilde{x}^{k(\delta)} - z\| = 0.$$

This follows from the evident estimate

$$\|A\tilde{x}^{k(\delta)} - y\| \le \|A\tilde{x}^{k(\delta)} - y_\delta\| + \|y_\delta - y_0\,| \le (1+\tau)\,\delta$$

and the known fact on continuity of the inverse mapping on a compact set for a one-to-one continuous operator A.

7 Ill-posed problems with *a priori* information

7.1 Formulation of the problem and convergence theorems

In the previous sections (see Sections 1–3) of this chapter, for the considered equations and systems of equations with additional constraints on the solution, conditions of uniqueness and, actually, well-posedness of the basic problem were discussed.

In this section we discuss the linear equation

$$Ax = y \tag{7.1}$$

with the bounded operator A acting on a pair of Hilbert spaces $\mathcal{H}_1, \mathcal{H}_2$ and without any assumptions about uniqueness of solution and boundedness of the inverse operator. Thus, we shall speak about an *essentially ill-posed problem* (Ivanov et al., 1978). The integral Fredholm equation of the first kind is a typical example of such a problem,

$$Ax \equiv \int_a^b K(t,s)x(s)\,ds = y(t), \quad c \le t \le d, \tag{7.2}$$

that is considered on the pair of spaces $L_2[a,b]$, $L_2[c,d]$ of the functions integrable in square. Equation (7.2) plays the role of a mathematical model of processes in various branches of the natural sciences in investigations of so called *inverse problems*; see (Tikhonov and Arsenin, 1979; Ivanov et al., 2002; Vasin and Ageev, 1995; Tikhonov et al., 1995; Bakushinskii and Goncharskii, 1994; Engl et al., 1996; Groetsch, 1993).

In the sequel, the intention is to describe more completely the mathematical formulation of an applied problem with the aim of obtaining high quality information about the object under investigation. The leads to the consideration (together with the basic equation) of additional relations and connections on the solution to be found. These additional constraints, represented in the form of equalities, inequalities or inclusions, describe important characteristics of the solution that are joined with its subtle structure and have not been revealed by the basic equation.

This usage of *a priori information* about the solution is especially important in considering the ill-posed formulation with a nonunique solution of the basic equation, since such an information allows to localize the solution to be found and select one which corresponds to the physical contents of the problem. It is also important to understand that in the case of nonuniqueness of the solution, neither method does guarantee obtaining the solution without additional information; this follows from the fact that, depending on the algorithm used and the initial data taken, we obtain different solutions, including those that do not describe the real object (phenomenon).

Thus equation (7.1) represents a mathematical model of the object under consideration and joins the solution x to be found with the initial data of the problem A, y, where the operator A is defined by physical principles (laws) used in the model description, and the element (function) y corresponds to the measured values (parameters).

Let some additional information on the solution of equation (7.1) be known in the form of inequalities

$$f_i(x) \le 0, \quad i = 1, 2, \dots, m, \tag{7.3}$$

where f_i are convex sub-differentiable (with bounded derivatives $\partial f_i(x)$), semi-continuous from below functionals given on the Hilbert space $\mathcal{H}_1$. System (7.3) reflects some structural characteristics of the solution that are not described by the basic equation (7.1).

It is necessary to find the solution of equation (7.1) which satisfies system (7.3). If we denote by $M \neq \varnothing$ the set of solutions of equation (7.1) and by Q the set of solutions of system (7.3), then we have to find an element $x \in M \cap Q$.

To solve this problem, it is natural to consider iterative processes in the form

$$x^{k+1} = P(U(x^k)), \tag{7.4}$$

$$x^{k+1} = \gamma\, P(x^k) + (1-\gamma)\, U(x^k), \quad 0 < \gamma < 1, \tag{7.5}$$

where U is the solving iterative operator for equation (7.1), and P is Q-pseudo-contractive or a Q-Fejér mapping that generates an iterative procedure for solving system (7.3).

If we take the step operator from one of the versions of the simple iteration method (see (2.6) and Remark 2.10, Chapter I) as the mapping U, i.e.,

$$U(x) = x - \beta\,(A^* Ax - A^* y), \quad \beta \le 1 / \|A\|^2, \tag{7.6}$$

or from the implicit iterative scheme, i.e., (see (2.11), Chapter I)

$$U(x) = (A^* A + \alpha\, I)^{-1}\,(\alpha\, x + A^* y), \quad \alpha > 0, \tag{7.7}$$

and if we define the mapping P by formulas (3.2)–(3.4) or (3.2), (3.3 a), (3.4), then we shall be able to satisfy the conditions of Theorem 3.9, Chapter I and obtain the following statement.

Theorem 7.1. *Under the given assumptions, the processes* (7.4) *and* (7.5) *generate the iterative sequence x^k that converges weakly to an element $x \in M \cap Q$.*

Proof. Firstly, ensure that the operator U defined by formula (7.6) belongs to the class $\mathcal{P}^1$ (see Definition 1.5, Chapter I), i.e., the following relation holds,

$$\|Ux - Uv\|^2 \le \|x - v\|^2 - \|Ux - x - (Uv - v)\|^2 \quad \forall\, x, v. \tag{7.8}$$

Note that (7.8) is equivalent to the following inequality,

$$\|Ux - Uv\|^2 \le (Ux - Uv,\ x - v) \quad \forall\, x, v.$$

Taking into account the linearity of the operator A and changing $x - v = u$, we obtain the latter inequality in the form

$$\|(I - \beta\, A^* A)\, u\|^2 \le ((I - \beta\, A^* A)u, u) \quad \forall\, u,$$

where the operator $T = I - \beta\, A^* A$ for $\beta \le \frac{1}{\|A\|^2}$ is positive semidefinite and selfadjoint with the norm

$$\|T\| = \sup_{\sigma \in \sigma(A^* A)} |I - \beta\sigma| \le 1.$$

Therefore there exists the operator $T^{1/2}$ with the norm $\|T^{1/2}\| \leq 1$. The chain of inequalities

$$\|Tu\|^2 = \|T^{1/2}T^{1/2}u\|^2 \leq \|T^{1/2}\|^2 \, \|T^{1/2}u\|^2$$
$$\leq \|T^{1/2}u\|^2 = \langle Tu, u\rangle$$

completes the proof of the inclusion $U \in \mathcal{P}^1$. Since it is evident that $\mathrm{Fix}(U) = M$, we have $U \in \mathcal{P}^1_M$, i.e., the operator U is M-pseudo-contractive.

By the same scheme, the inclusion $U \in \mathcal{P}^1$ is proved for the operator U from (7.7).

By virtue of linearity of the operator A, property (3.1) from Chapter I is evidently satisfied, and for the operator P this holds because of Remark 3.15, Chapter I. Making use of Theorem 3.1, Chapter I, completes the proof. □

Remark 7.2. In the theorem it is possible to take $0 < \beta < 2/\|A\|^2$ (see (Vasin and Ageev, 1995)).

To obtain a strongly converging sequence of iterations, use the method of correcting multipliers and construct the iterative schemes

$$x^{k+1} = \gamma_{k+1} P(U(x^k)) + (1 - \gamma_{k+1})\, v_0, \tag{7.9}$$

$$x^{k+1} = \gamma_{k+1}[\gamma\, P(x^k) + (1 - \gamma)U(x^k)] + (1 - \gamma_{k+1})\, v_0. \tag{7.10}$$

Now let the *a priori* constraints (7.3) are given in the form of the system of linear inequalities

$$l_j(x) \equiv (a_j, x) - b_j \leq 0, \quad j = 1, 2, \ldots, m. \tag{7.11}$$

Define the mapping

$$P_j(x) = x - \lambda \frac{l^+(x)}{\|a_j\|^2} a_j, \quad \lambda \in (0, 2), \;\; a_j \neq 0,$$

and construct the operators

$$P^{(1)}(x) = \sum_{i=1}^{m} \alpha_i P_j(x), \quad \alpha_i \in (0, 1), \; \sum_{i=1}^{m} \alpha_i = 1, \tag{7.12}$$

$$P^{(2)}(x) = P_{j_1} P_{j_2} \ldots P_{j_m}(x). \tag{7.13}$$

Note that since the mapping P_j is nonexpansive (see Lemma 3.1, Chapter IV), the mappings $P^{(1)}$ and $P^{(2)}$ will also be nonexpansive. As it has been established in Theorem 7.1, the operators U defined in (7.6), (7.7) are pseudo-contractive (of the class $\mathcal{P}^1$) and, therefore, nonexpansive.

Gathering these facts, we obtain the following corollary from Theorem 4.5, Chapter I (here, the ball with the center at the point $\bar{x} \in M \cap Q$ plays the role of the set $\mathcal{D}$).

Corollary 7.3. *For any admissible sequence γ_k (see Definition 4.4, Chapter I) and any element $v_0 \in \mathcal{H}_1$, the iterations x^k generated by processes* (7.9) *and* (7.10) *strongly converge to the element $x \in M \cap Q$ closest to v_0, where the operator $P^{(1)}$ or $P^{(2)}$ from* (7.12), (7.13) *take the role of the operator P, resp.* □

7.2 Properties of iterations under noisy data

As it was noted in Section 4 (Subsection 4.3), Chapter I, for the iterative approximation of a solution of an ill-posed problem with noisy data, the necessary number of iterations has to be related with the given error in the data. There, a rule for the choice of such a dependence was also formulated for the method of correcting multipliers.

We illustrate this rule for process (7.9) when the mapping $P^{(1)}$ is used as the operator P.

Let the right-hand side y of equation (7.1) and the vector $b = (b_1, b_2, \dots, b_m)^\top$ of the free terms in the inequality system (7.11) be given approximately by the pair $\tilde{y}, \tilde{b}$, such that

$$\|y - \tilde{y}\| \le \delta, \qquad \max_{1 \le i \le m} |b_i - \tilde{b}_i| \le \delta.$$

Introduce the following notations:

$$\tilde{l}_j(x) = (a_j, x) - \tilde{b}_j, \quad \widetilde{P}_j(x) = x - \lambda \frac{\tilde{l}^+(x)}{\|a_j\|^2} a_j.$$

Then we have the estimates

$$\|P_j(x) - \widetilde{P}_j(x)\| = \frac{\lambda}{\|a_j\|} |l^+(x) - \tilde{l}^+(x)| \le \frac{\lambda}{\|a_j\|} \delta,$$

$$\|P^{(1)}(x) - \widetilde{P}^{(1)}(x)\| = \sum_{j=1}^{m} \alpha_j \|P_j(x) - \widetilde{P}_j(x)\| \le \frac{\lambda}{\|a\|} \delta,$$

where $\|a\| = \min_{1 \le j \le m} \|a_j\|$.

For the operators U given by formulas (7.6) and (7.7), we find, correspondingly,

$$\|U(x) - \widetilde{U}(x)\| = \beta \, \|A^* y - A^* \tilde{y}\| \le \beta \, \|A^*\| \, \delta,$$
$$\|U(x) - \widetilde{U}(x)\| = \|(A^* A + \alpha \, I)^{-1} \, A^*(y - \tilde{y})\| \le \frac{\|A^*\|}{\alpha} \, \delta.$$

Gathering the obtained results, we have for the operator $T = \gamma \, P^{(1)} + (1 - \gamma) \, U$ the following estimate uniform on x,

$$\|T(x) - \widetilde{T}(x)\| \le \varphi(\delta),$$

where either $\varphi(\delta) = (\gamma\lambda \,/\, \|a\| + (1-\gamma)\, \beta \, \|A^*\|) \, \delta$, if U is given accordingly to (7.6), or $\varphi(\delta) = (\gamma\lambda \,/\, \|a\| + (1-\gamma) \, \|A^*\| \,/\, \alpha) \, \delta$, if U satisfies (7.7).

Thus, a condition for the approximation of the operator T is fulfilled that is stronger than each of the conditions (4.9), (4.10), Chapter I. On the basis of Theorem 4.6, Chapter I, and under relation $k(\delta) \cdot \varphi(\delta) \to 0$ as $\delta \to 0$ of the parameters, this guarantees strong convergence of the sequence $\tilde{x}^{k(\delta)}$ generated by process (7.10) to the solution $\hat{x} \in M \cap Q$, with $P = P^{(1)}$ and y, b are substituted by their approximations $\tilde{y}$, $\tilde{b}$ with the level δ of the error.

In the more general situation when the vectors a_j and the operator A are given with the error $\|a_j - \tilde{a}_j\| \le \delta$, $\|A - \widetilde{A}\| \le \delta$, after simple calculations and for sufficiently small δ, we obtain

$$\|P_j(x) - \widetilde{P}_j(x)\| \le \frac{\lambda \, (1 + \|x\|)}{\|a\|} \, \delta,$$
$$\|U(x) - \widetilde{U}(x)\| \le \Big[\beta \, (\|A^*\| + \|\widetilde{A}\|) \, \|x\| + \beta \, (\|y\| + \|\widetilde{A}^*\|)\Big] \, \delta$$

for the operator U from (7.6), and

$$\|U(x) - \widetilde{U}(x)\| \le \Big[h \, \|x\| + (\|A^*\| + \|y\|) \,/\, \alpha + h \, \|\widetilde{A}^*\| \, \|\tilde{y}\|\Big] \, \delta,$$

where $h = (\|A\| + \|\widetilde{A}^*\|) \,/\, (\alpha^2 - \alpha \, (\|A\| + \|\widetilde{A}^*\|)) \, \delta$, for the operator U from (7.7).

The obtained estimates and the boundedness of x^k (i.e., the iterations for exact data) allow to conclude that the condition of approximation (4.10), Chapter I, holds for the operator $T = \gamma \, P^{(1)} + (1-\gamma) U$. As above under the relation $k(\delta) \cdot \varphi(\delta) \to 0$, this guarantees convergence of process (7.10) in the case of noisy data (here, $\varphi(\delta)$ is easily written out from the presented estimates).

7.3 Taking into account the *a priori* information in applied problems

1. In the investigation of the atomic structure of one-component unordered (amorphous) materials by the EXAFS (Extended X-ray Absorption of Fine Structure), there appears the integral *Fredholm equation of the first kind* (Vasin and Ageev, 1995, and the references therein)

$$Ag \equiv \int_a^b K(k,r)g(r)\,dr = \chi(k), \quad c \le k \le d, \tag{7.14}$$

where $\chi(k)$ is the absorption coefficient of the monochrome X-ray beam in the material under study, which is measured experimentally; $g(r)$ is the radial atomic distribution function (to be found) that is the most important structural characteristic of the material. From the physical sense of the function $g(r)$, it follows

$$g(r) \ge 0, \tag{7.15}$$

$$(g,v) \equiv \frac{3}{b^3 - a^3} \int_a^b r^2 g(r)\,dr = 1. \tag{7.16}$$

Theoretical and numerical analysis showed that the integral operator in (7.14) has a non-trivial kernel, and, hence, the solution of the equation is not unique.

The approximate regularized solution obtained by the classical Tikhonov method (without using the *a priori* information (7.15), (7.16)), has no physical sense. This means that the normal solution, which is approximated by this method, does not satisfy conditions (7.15), (7.16).

To obtain a solution having physical sense, two approaches have been suggested. In the first one, the *iterated version of the Tikhonov regularization* is applied together with projecting onto *a priori* sets. For this purpose, the following mapping is introduced,

$$U : x \to \arg\min_g \{\|Ag - \chi\|^2_{L_2} + \alpha\,\|g - x\|^2_{W_2^1 \text{ (or } L_2)}\}$$

and the following iterative process is constructed,

$$g^{k+1} = P_{Q_2} P_{Q_1} U(g^k). \tag{7.17}$$

Here, P_{Q_1} is the metric projection onto the set $Q_1 = \{g : \langle g,v\rangle = 1\}$, and P_{Q_2} is the projection onto the set $Q_2 = \{g : g \ge 0\}$.

In the second approach, the normalization condition (7.16) is taken into account by addition of this equation to the system obtained in the result of discrete approximation of equation (7.14), and the second condition (7.15) is similar to the one shown above in the first approach.

Numerical experiments for equation (7.14) for simulated and experimental data have shown that an application of the *a priori* constraints by the suggested procedure in the algorithm allows to obtain a solution of reasonable quality. The details of the numerical simulation can be found in (Vasin and Ageev, 1995).

2. Under violation of the monotonic dependence of the electron concentration as the function of altitude, the so called problem of waveguides appears in the oblique radiolocation of the ionosphere from the ground surface when the vertical profile of the electron concentration has to be defined along the epicentral distances (as the function of the beam parameter) obtained on two frequencies.

This is related to the fact that the solution of the problem becomes nonunique in this case and, moreover, the traditional method allows to define the concentration only until beginning on the waveguide (i.e., until monotonicity is violated). In the work (Ageev et al., 1997), it was shown that in spite of the impossibility to define the concentration on the waveguide in the one-valued way, one can determine the "waveguide measure" (i.e., the measure of the Lebesgue sets for the wave refraction coefficient) if we solve the corresponding *Fredholm–Stieltjes equation* of the form

$$\int_a^b K(p,r)\,dF(r) = R(p), \tag{7.18}$$

where $K(p,r)$ is some given function, $R(p)$ is a function defined experimentally, and $F(r)$ is the waveguide measure. Moreover, the electron concentration over the waveguide can be calculated by means of the determined function $F(r)$.

When the Tikhonov regularization is used in the standard form, the quality of the solution appears to be not very well. If we additionally take into account the *a priori* data about the solution F in the form

$$F(a) = 0, \quad F'(b) = 0, \quad F'(r) \geq 0, \quad F(r) \geq 0, \tag{7.19}$$

then applying the iterative process of the form (7.17) enhances essentially the solution quality. In this case $P_{Q_1} = P_1 P_2 \dots P_n$ or $P_{Q_1} = \sum_{i=1}^n \lambda_i P_i$, where P_i is the projection onto the half-space (see formula (3.7)) that is defined by the ith inequality in the system $BF \geq 0$ with the two-diagonal matrix B; this matrix is obtained after finite-difference approximation of the condition $F'(r) \geq 0$. The operator P_{Q_2} is the metric projection onto the set Q_2, which is defined by the condition $F(x) \geq 0$, i.e., $P_{Q_1}F = F^+$ is the positive cut-off-function of the vector F.

Note that after integration by parts of equation (7.18) and taking into account the boundary conditions from (7.19), we obtain the usual Fredholm equation

$$\int_a^b K'_r(p,r)F(r)\,dr = -R(p), \tag{7.20}$$

where it is necessary to apply an iterative scheme.

We next describe another possible way for constructing iterative processes for solving equation (7.20). Consider some grid $\{r_i\}$, $i = 0, 1, \dots, m+1$, and approximate the condition $F'(r) \geq 0$ by the system of inequalities

$$F(r_{i+1}) - F(r_i) \geq 0, \quad i = 0, \dots, m, \tag{7.21}$$

where $F(r_0) = F(a) = 0$, $F(r_{m+1}) = F(r_m) = 0$. Introducing the artificial variables Y_i $(i = 1, \dots, m)$, we pass from the inequality system (7.21) to the system of linear algebraic equations

$$\begin{aligned} F_1 + 0 + 0 + \cdots + 0 - Y_1 &= 0, \\ -F_1 + F_2 + 0 + \cdots + 0 - Y_2 &= 0, \\ &\vdots \\ 0 + 0 + 0 \cdots - F_{m-1} + F_m + 0 + \cdots + 0 - Y_m &= 0, \end{aligned} \tag{7.22}$$

for which we search the solution under the condition

$$Z = (F_1, F_2, \dots, F_m, Y_1, Y_2, \dots, Y_m) \geq 0.$$

In matrix notation, the system (7.22) takes the form

$$CZ = 0, \quad Z = [F, Y]^\top. \tag{7.23}$$

Denote by A the matrix obtained as a result of discretization of the integral operator in problem (7.20), and denote by b the vector that approximates the right-hand side. Then the system of linear algebraic equations approximating problem (7.20) can be formally written in the form

$$\mathcal{D}z = d,$$

where $\mathcal{D} = [A, \boldsymbol{O}]$, $\boldsymbol{O}$ is the zero matrix, $d = (b, 0)^\top$.

Thus, the original problem is reduced to finding the nonnegative solution $z = (p, v)^\top \geq 0$ of the compound system

$$\begin{cases} \mathcal{D} z = d, \\ C z = 0. \end{cases}$$

For the approximation of its solution, it is possible to use a broad range of methods of the form

$$z^{k+1} = P^+ V(z^k),$$

where, for instance, V is the step operator of some explicit or implicit scheme (see (7.6), (7.7)) with the matrix $B = [\,\mathcal{D}, C\,]^\top$, and P^+ is a projection onto the positive orthant.

3. A problem of reconstruction an image corrupted by the hardware of the measuring equipment and by an additive noise is reduced to solving the two-dimensional integral equation of the first kind

$$Au \equiv \int_0^1 \int_0^1 K(x - t, y - s)\, u(t, s)\, dt\, ds = f(x, y), \tag{7.24}$$

this is a typical ill-posed problem. Thus, for the stable construction of an approximate solution, it is necessary to use ideas of regularization, for example, on the basis of the *Tikhonov method*

$$\min \{\|Au - f\|^2 + \alpha\, \Omega(u - u^0) : u \in U\}. \tag{7.25}$$

Numerical experiments, see, for instance, (Vogel, 2002), show that for problems of type (7.24) originating from the reconstruction of images, applications of the Tikhonov stabilizers in the form

$$\Omega(u) = \|u\|^2_{W_2^l[\Pi]}, \qquad \Pi = \{(x, y) : 0 \leq x, y \leq 1\}$$

with the norm of the Sobolev space ($l \geq 1$) is not suitable in many cases. This is caused by the effect of smoothing of a discontinuous solution. Application of the stabilizing functional of the form

$$\Omega(u) = \|u\|^2_{L_2} + J(u) \tag{7.26}$$

with the variation $J(u)$ of this or that type, see (Leonov, 1996; Acar and Vogel, 1994; Vasin, 2002), as a rule, leads to better results. In particular, the following total variation (Giusti, 1984) can be used as the functional $J(u)$,

$$J(u) = \sup \left\{ \int_0^1 \int_0^1 u(x, y) \operatorname{div} v(x, y)\, dx\, dy : v \in C_0^1(\Pi, R^2),\ |v| \leq 1 \right\},$$

which in the case of a smooth function $u \in W_1^1(\Pi)$ takes the form

$$J(u) = \int_0^1 \int_0^1 |\nabla u| \, dx \, dy.$$

In the general multi-dimensional case, a theorem on convergence (Vasin, 2002) for the Tikhonov method (7.25) with stabilizer (7.26) is proved, namely,

$$\lim_{\delta \to 0} \|u^{\alpha(\delta)} - \hat{u}\|_{L_2(\Pi)} = 0, \quad \lim_{\delta \to 0} J(u^{\alpha(\delta)} - u^0) = J(\hat{u} - u^0),$$

with the relation of the parameter $\delta / \alpha(\delta) \to 0$, $\alpha(\delta) \to 0$, where u^α is the solution of the extremal problem (7.25); $\hat{u}$ is the normal solution of the integral equation; δ is the level of error in the input data.

To reduce problem (7.25) to a finite-dimensional problem, the discretization of the integral operator A and the stabilizer Ω is implemented on the basis of the two-dimensional analogue of the rectangular formula. After some transformations (summation by parts), the finite-dimensional extremal problem is transformed to the form

$$\min \Big\{ \sum_{k,l} \Big[\sum_{i,j} h^2 K(x_k - t_i, y_l - s_j) u(t_i, s_j) - f_{kl} \Big]^2 h^2 \\ + \alpha \Big\{ \sum_{i,j} h^2 u^2(t_i, s_j) + \sum_{i,j} h^2 \Big[\Big(\frac{u_{i,j} - u_{i,j-1}}{h} \Big)^2 \\ + \Big(\frac{u_{i,j} - u_{i-1,j}}{h} \Big)^2 \Big] \Big\}^{\frac{1}{2}} : u_n \in l_2^n \Big\}. \quad (7.27)$$

Denote the objective functional in problem (7.27) by $\Phi_n^\alpha(u_n)$, and denote its optimal value by Φ_n^*. We assume that some estimate $\widetilde{\Phi}_n$ of Φ_n^* is known, i.e., $\widetilde{\Phi}_n \le \Phi_n^* + \varepsilon$, $\varepsilon > 0$. Now the problem of approximate minimization (7.27) can be written as the solution of the inequality

$$\Phi_n^\alpha(u_n) \le \widetilde{\Phi}_n \quad (7.28)$$

with a convex function of n variables. Denote by M the set of solutions of inequality (7.28).

Since the stabilizing term in $\Phi_n^\alpha(u_n)$ is not differentiable, it is natural to use some subgradient method for solving equation (7.28). Taking into account an *a priori* information of the form

$$Q = \{u : u \ge 0\}$$

about the solution, we consider the iterative process

$$u_n^{k+1} = P^+\Big\{u_n^k - \frac{[\Phi_n^\alpha(u_n^k) - \widetilde{\Phi}_n]^+ \, \partial\Phi_n^\alpha(u_n^k)}{\|\partial\Phi_n^\alpha(u_n^k)\|^2}\Big\}, \tag{7.29}$$

where $\partial\Phi_n^\alpha(u_n^k)$ is an arbitrary subgradient of the function $\Phi_n^\alpha(u)$ at the point u_n^k, and P^+ is the operator of the positive cut-off-function (i.e., the projection operator onto the set Q).

The step operator of process (7.29) represents the superposition of two operators from the classes $\mathcal{P}_M^1$ and $\mathcal{P}_Q^1$ (see Lemmas 3.13 and 3.14, Chapter I) that on the basis of Theorems 3.1 and 6.10, Chapter I, guarantees convergence of the iterations to some solution $\hat{u} \in M \cap Q$.

In the work (Vasin and Serezhnikova, 2004) there is a detailed description of the numerical results obtained by implementation of algorithm (7.29) on the reconstruction of a corrupted image for the real data K, f taken from the internet (Vogel and Hanke, 1998) in the form of numerical arrays on a 128×128 grid in the square $0 \le x, y \le 1$.

This approach revealed to be rather efficient and allows to select more legibly the contours of the image. Comparison of the reconstructed image with the results obtained in the work (Leonov, 1999) where another approach was used (this one is based on the preliminary smoothing for the nonsmooth stabilizer and further piecewise- uniform approximation) shows their vicinity in quality.

4. For the investigation and interpretation of productivity of a wellbore/ reservoir system, the problem of the solution of the convolution Volterra equation (Schroter et al., 2001) arises:

$$\int_0^t q(t-\tau)g(\tau)d\tau = \Delta p(t), \quad \Delta p(t) = p_0 - p(t), \;\; t \in [0, T]. \tag{7.30}$$

Here, $p(t)$ and $q(t)$ are the pressure and the flow rate, respectively, p_0 is the initial reservoir pressure, and $g(t) = dp_u(t)/dt$, where $p_u(t)$ is the so called constant-unit-rate pressure response. The functions $p_u(t)$ and $g(t)$ are unknown and are to be reconstructed by $\Delta p(t)$ and $q(t)$ that are given with noise. The functions $p_u(t)$ and $g(t)$ are used for the analysis of a wellbore/ reservoir system (Bourdet et al., 1989).

As it is known (Coats et al., 1964), the function $g(t)$ satisfies the following constraints:

$$C \geqslant g(t) \geqslant 0, \quad g(t) \leqslant 0, \quad g'(t) \leqslant 0, \quad g''(t) \geqslant 0. \tag{7.31}$$

So, we must solve the system

$$Ag = \Delta p, \quad g \in M, \tag{7.32}$$

where $M = \{g : 0 \leqslant g \leqslant C,\ g'(t) \leqslant 0,\ g''(t) \geqslant 0\}$.

Now the following iterative processes

either

$$g^{k+1} = P_M(g^k - \beta(A^*Ag^k - A^*\Delta p)), \quad 0 \leqslant \beta \leqslant 2/\|A\|^2, \tag{7.33}$$

or

$$g^{k+1} = P_M(A^*A + \alpha I)^{-1}(\alpha g^k + A^*\Delta p)$$

are appropriate for solving system (7.32). Here, the mapping P_M is given, for example, by the formulas (7.12), (7.13) or (3.12), (3.13) from Chapter I.

Since the set of functions M given by inequalities (7.31) is compact in $L_2[0, T]$, we can also use the Ivanov quasi-solution method (Ivanov et al., 2002)

$$\min\{\|Ag - \Delta p\|^2 : g \in M\}. \tag{7.34}$$

After discretization of problem (7.34) by a finite-difference method, we arrive at the minimization of a quadratic function with linear constraints, which can be solved by methods of the gradient type.

Another approach for solving equation (7.30) is based on the preliminary transition from the linear equation (7.30) to a nonlinear one,

$$A(z) = \int_{-\infty}^{\log_{10} t} q(t - 10^{\sigma})\, 10^{\sigma}\, d\sigma = \Delta p(t), \quad t \in [0, T] \tag{7.35}$$

after changing of the variable τ and the function $g(t)$ as follows:

$$\sigma = \log_{10} \tau, \quad z(\sigma) = \log_{10}(\tau g(\tau)).$$

Now for solving (7.35), we can apply an iterative processes of the form

$$g^{k+1} = P_M V(g^k),$$

where V is the step operator of the iterative method of Gauss–Newton type (e.g., (6.4) in Chapter II), and P_M is a Fejér mapping corresponding to the constraints (7.31) (see (1.1)–(1.5), Chapter III).

5. In the inverse problems of thermic atmosphere, it is necessary to find the temperature $T(h)$ and concentration $n(h)$ of the green-house gases (CO, CO_2, CH_4, etc.) as function of the height using spectra measured by the satellite sensors. Here, the following integral equation arises,

$$A(\nu) = \int_{-\infty}^{+\infty} W(\nu) F(\nu - \nu') d\nu' = \Phi(\nu). \tag{7.36}$$

This equation is the convolution of a spectrum W of high resolution measured with apparatus function F. The function W depends on the absorption coefficient $B_\nu(T(h))$ that is a nonlinear function of unknown parameters $u(h) = (T(h), n(h))$.

Usually, only a part of parameters is found (i.e., the temperature and methane; the temperature, water vapor, and carbon acid), but other parameters are considered to be fixed and their values are chosen from the database for the region under investigation.

Peculiarity of such a problem is in the presence of *a priori* information for the unknown solution in form of the two-sided inequalities

$$Q = \{u : \underline{u}(h) \leqslant u(h) \leqslant \overline{u}(h)\}. \tag{7.37}$$

For the numerical reconstruction of a solution of equation (7.34), satisfactory results are obtained by using an iterative method of the regularized Gauss–Newton type, in particular, the modified Levenberg–Marquardt method (6.4):

$$u^{k+1} = P_Q[u^k - \beta(A'(u^k)^* A'(u^k) + \alpha_k I)^{-1} A'(u^k)^* (A'(u^k) - \Phi)]$$

where P_Q is the projection operator on the n-dimensional parallelepiped (7.5) (note that after discretization, $u(h)$ is an n-dimensional vector) (see (Vasin et al., 2006)).

Chapter III

Application of Fejér methods to solve linear and convex inequalities

1 The basic construction of M-Fejér mappings for application to finite systems of linear inequalities

We consider the following system of linear inequalities in a Hilbert space $\mathcal{H}$:

$$l_j(x) = (a_j, x) - b_j \le 0, \quad j = 1, \dots, m. \tag{1.1}$$

We shall suppose that $a_j \neq 0$ for all j. The projection $P_{Q_j}(x)$ of the point x onto the half-space $Q_j = \{x : l_j(x) \le 0\}$ with the relaxation coefficient $\lambda_j \in (0, 2)$ is defined by the formula

$$T_j(x) = P_{Q_j}(x) = x - \lambda_j \frac{l_j^+(x)}{\|a_j\|^2} \cdot a_j. \tag{1.2}$$

The validity of the inclusion $T_j \in \mathcal{F}_{Q_j}$ was considered in Section 5, Chapter I. Assume $M = \bigcap_{j=1}^m Q_j \neq \varnothing$. We consider now three *basic constructions* of M-Fejér mappings.

1. *Weighted projection* construction:

$$T^{(1)}(x) = \sum_{j=1}^m \alpha_j T_j(x) = x - \sum_{j=1}^m \alpha_j \lambda_j \frac{l_j^+(x)}{\|a_j\|^2} \cdot a_j. \tag{1.3}$$

Here, $\alpha_j > 0$, $\sum_{j=1}^m \alpha_j = 1$.

2. *Cyclic projection* construction:

$$T^{(2)}(x) = T_{j_1} T_{j_2} \dots T_{j_m}(x), \tag{1.4}$$

where $(j_1, \dots, j_m)$ is an arbitrary ordering of the indices $j = 1, \dots, m$.

3. *Extremal projection* construction:

$$T^{(3)}(x) = \left\{ x - \lambda \frac{d^+(x)}{\|a_{j_x}\|^2} \cdot a_{j_x} : j_x \in J(x) \right\}, \tag{1.5}$$

$$\lambda \in (0, 2), \quad d(x) = \max_j l_j(x), \quad J(x) = \{j : d(x) = l_j(x)\}.$$

The mapping $T^{(3)}(x)$ is a contraction of mapping (7.6), Chapter I, which is closed and satisfies the statements of Corollary 7.7, Chapter I. The mappings $T^{(1)}$ and $T^{(2)}$ are continuous. Since any process $x^{k+1} \in T^i(x^k)$, $i = 1, 2, 3$ for initial x_0 operates in the finite-dimensional subspace $\mathrm{Lin}\{a_j^\top, x_0\}$ of the space $\mathcal{H}$, the convergence of all such sequences to the element $x' \in M$ is provided on the basis of Theorem 6.10, Chapter I. Thus, the following theorem can be formulated.

Theorem 1.1. *The sequence $\{x^k\}$, generated by any of the mappings $T^{(i)}(x)$, $i = 1, 2, 3$ and for arbitrary initial $x_0 \in \mathcal{H}$, converges to some solution of the consistent system* (1.1). □

When constructing the mapping $T^{(1)}$, the weight coefficients α_j can be chosen in various ways; for example, it is possible to put $\alpha_j = \frac{1}{m}$, $\alpha_j = \frac{\|a_j\|^2}{\sum_{s=1}^m \|a_s\|^2}$, and so on. They also can be variable, for instance, $\alpha_j(x) = \frac{l_j^+(x)}{\sum_{s=1}^m l_s^+(x)}$. In the latter case in definition of the mapping $T^{(1)}$, it is necessary to make small change, namely,

$$T^{(1)_0}(x) = \begin{cases} \sum_{j=1}^m \alpha_j(x) T_j(x), & x \notin M, \\ x, & x \in M. \end{cases} \tag{1.6}$$

Note that in (1.6) we have $x \in M \sim \delta(x) = \sum_{j=1}^m l_j^+(x) = 0$ and $x \notin M \sim \delta(x) > 0$; here, "$\sim$" is the symbol of equivalency.

It is possible to check the closedness of the mapping (1.6) that implies the validity of Theorem 1.1 for this mapping, i.e., the sequence x_k defined accordingly to the inclusion $x_{k+1} \in T^{(1)_0}(x_k)$ converges to some element $x' \in M$.

2 Fejér processes with variable coefficient of relaxation

For the case $\mathcal{H} = \mathbb{R}^n$, consider the process

$$x^{k+1} \in T_k(x^k), \quad x^0 \in \mathbb{R}^n, \tag{2.1}$$

where

$$T_k(x) = \left\{ x - \lambda_k \frac{d^+(x)}{\|h\|^2} h : h \in \partial d(x) \right\}, \tag{2.2}$$

with $\lambda_k \in [\varepsilon, 2 - \varepsilon]$, $\varepsilon \in (0, 1)$, $d(x)$ being a convex function, and finally $M = \{x : d(x) \leq 0\} \neq \varnothing$. The mapping (2.2) corresponds to relation (7.3), Chapter I, for $\lambda = \lambda_k$ and $E(x) = \partial d(x)$.

2.1 The main theorem

Theorem 2.1. *Under the conditions mentioned above, the process* (2.1)–(2.2) *converges to* $x' \in M$.

Proof. Let us to carry out the proof by a new scheme different from the one of Theorem 7.6, Chapter I.

1. If for some $x^{\bar{k}}$ we have $d^+(x^{\bar{k}}) = 0$, i.e., $x^{\bar{k}} \in M$, then the sequence x^k is stabilized, and, hence, the theorem is valid.

2. Let $\{x^k\} \cap M = \varnothing$ and $\{x^k\}' \cap M \neq \varnothing$, i.e., the sequence x^k contains a subsequence $x^{j_k} \to x' \in M$; above, $\{x^k\}'$ was the set of all limit points of the sequence x^k. Since in the situation under consideration x^k will be M-Fejér and $x' \in M$, we have $x^k \to x'$ according to Corollary 5.5, Chapter I. This completes the proof.

3. The case $\{x^k\}' \cap M = \varnothing$ corresponds to the fact that $\underline{\lim}_{k\to\infty} d(x^k) = \gamma > 0$. Actually, if $\gamma = 0$, then it is possible to select a subsequence x^{j_k} converging, say, to x', such that $\lim d(x^{j_k}) = d(x') = 0$, and this gives $x' \in M$ in contradiction to $\{x^k\}' \cap M = \varnothing$. Thus, $\gamma > 0$. Rewrite process (2.1) in the transformed form

$$x^{k+1} = x^k - \underbrace{\left[\lambda_k \frac{d(x^k)}{d_\delta(x^k)}\right]}_{\lambda'_k} \cdot \frac{d_\delta(x^k)}{\|h^k\|^2} \cdot h^k, \quad h^k \in \partial d(x^k), \tag{2.3}$$

where $d_\delta(x^k) = d(x^k) - \delta$, $0 < \delta < \gamma$. Note that $\partial d(x^k) = \partial d_\delta(x^k)$ for any δ. Since $d(x^k) \geq \gamma_1 = \gamma - \varepsilon_1 > 0$ for large k and sufficiently small $\varepsilon_1 > 0$, we can find the following estimate for λ'_k:

$$\lambda'_k = \lambda_k \frac{d(x^k)}{d(x^k) - \delta} = \lambda_k \left[1 + \frac{\delta}{d(x^k) - \delta}\right] \leq (2 - \varepsilon)\left[1 + \frac{\delta}{\gamma_1 - \delta}\right].$$

So as $1 + \frac{\delta}{\gamma_1 - \delta} \to 1$ for $\delta \to 0$, then for sufficiently small $\delta > 0$ there exists an $\bar{\varepsilon} > 0$ such that $\lambda'_k \leq 2 - \bar{\varepsilon}$, i.e., $\varepsilon \leq \lambda_k \leq \lambda'_k \leq 2 - \bar{\varepsilon}$ for all $k = 1, 2, \ldots$. Thus, the sequence x^k can be regarded as a result of a scheme of type (2.1), but in application to the mapping

$$W_k(x) = \left\{ x - \lambda'_k \frac{d_\delta^+(x)}{\|h\|} \cdot h : \, h \in \partial d_\delta(x) \right\}, \tag{2.4}$$

the previous h^k from $\partial d(x^k)$ are taken for the generation. But since $\partial d(x^k) = \partial_\delta d(x^k)$, the previous choice of h^k will also correspond to rule (3.3). Put $M_\delta = \{x : d_\delta(x) = d(x) - \delta \le 0\}$. The set $M = \{x : d(x) \le 0\}$ will be a part of the interior of the set M_δ, therefore, M_δ is solid, i.e., it has an nonempty interior. So as the sequence x^k will be M_δ-Fejér, then accordingly to Corollary 5.4, Chapter I, this sequence has a unique limit point, i.e., $x^k \to x'$. We now show that $x' \in M$. Rewrite the realization of process (2.1) under consideration in the form

$$x^{k+1} = x^k - \lambda_k \frac{d^+(x^k)}{\|h^k\|^2} \cdot h^k. \tag{2.5}$$

From (2.5) it follows

$$d^+(x^k) = (\|h^k\| / \lambda_k) \, \|x^{k+1} - x^k\|. \tag{2.6}$$

Since $\sup_k \|h^k\| < +\infty$ (see Lemma 7.3, Chapter I) and $\lambda_k \ge \varepsilon$, it follows from (2.6) that $\lim d^+(x^k) = d^+(x') = 0$, i.e., $x' \in M$.

The theorem is completely proved. □

Remark 2.2. Sometimes it could be useful to determine the relaxing coefficients λ_k according to the rule $\lambda_{k+1} = \lambda\,(x^k)$, where $\lambda(x)$ is some function that maps $\mathbb{R}^n$ onto the segment $[\varepsilon, 2 - \varepsilon]$. The choice of λ_k may also be arbitrary. The validity of Theorem 2.1 is not violated here.

2.2 Fejér process with the mirror relaxation

In application to system (1.1), consider the mapping (1.5) for $\lambda = 2$ and the generated sequence x^k described by the following recurrent relation:

$$x^{k+1} = x^k - 2\frac{d^+(x^k)}{\|a_{j_k}\|^2} \cdot a_{j_k}, \quad j_k \in J(x^k), \tag{2.7}$$

where $J(x) = \{j : d(x) = l_j(x)\}$. In (2.7), the value $\lambda = 2$ is taken (that earlier was excepted) as the relaxation coefficient which is used in all constructions of M-Fejér mappings for $\lambda \in (0, 2)$. This version is called the *mirror relaxation*. In this case process (2.7) will be weakly M-Fejér in the following sense:

$$\forall\, y \in M : \ \|x^{k+1} - y\| \le \|x^k - y\| \quad \forall\, k, \tag{2.8}$$

but inequalities (2.8) will be strict for $x^k \notin M$ if $y \in M^0$; here, for M^0, the interiority of the polyhedron M is understood under the assumption that it is solid.

Process (2.7) is interesting due to the fact that under such an assumption about the polyhedron M, this process terminates, i.e.,

$$\exists \bar{k} : x^{\bar{k}} \in M.$$

We shall need one simple identity that in essence was already used in Chapter I. Namely, let Q be a half-space of a Hilbert space $\mathcal{H}$ given by the inequality $l(x) = (a, x) - \alpha \leq 0$, $a \neq 0$. Define the mapping

$$P_Q^\lambda(x) = x - \lambda \frac{l(x)}{\|a\|^2} \cdot a, \quad \lambda \in (0, 2].$$

If $x \notin Q$, then $P_Q^\lambda(x)$ is the projection operator onto the half-space Q with the relaxation λ.

Lemma 2.3. *The following identity is valid for the mapping* P_Q^λ*:*

$$\| P_Q^\lambda(x) - y\|^2 = \|x - y\|^2 + 2\lambda \frac{l(x)l(y)}{\|a\|^2} - \lambda(2 - \lambda) \frac{[l(x)]^2}{\|a\|^2}$$

with x *and* y *from* $\mathcal{H}$.

Proof. The statement is checked out directly. □

The following theorem holds.

Theorem 2.4. *If the polyhedron* M *of system* (1.1) *is solid, then process* (2.7) *terminates, i.e., at some finite step of the process we obtain a point from* M.

Proof. This fact is proved by contradiction, i.e., we assume $\{x^k\} \cap M = \varnothing$. Immediately note that the step $x^k \to x^{k+1}$ (according to (2.7)) is in the mirror mapping of the point x^k with respect to the hyperplane $H_k = \{x : (a_{j_k}, x) - b_{j_k} = 0\}$; so, if $y \in H_k$, then

$$\|x^{k+1} - y\| = \|x^k - y\|. \tag{2.9}$$

Here $j_k \in J(x^k)$.

According to the identity from Lemma 2.3 above applied to (2.7), we have

$$\|x^{k+1} - y\|^2 = \|x^k - y\|^2 + 4 \frac{l_{j_k}(x^k)\, l_{j_k}(y)}{\|a_{j_k}\|^2}, \quad y \in M. \tag{2.10}$$

From this relation, for $y \in M^0 = \operatorname{int} M$ (in this case $l_{j_k}(y) < 0$) it follows

$$\|x^{k+1} - y\| < \|x^k - y\|, \tag{2.11}$$

i.e., the sequence x^k is M^0-Fejér, which by the assumption that M^0 is solid gives convergence of the sequence x^k (and the same fact follows from the weak M-Fejér property of this sequence). Therefore, $x^k \to x'$. Let us prove that $x' \in M$. It follows from (2.7), namely,

$$\begin{aligned} d(x^k) = (1/2)\,\|x^{k+1} - x^k\| \cdot \|a_{j_k}\| \to 0 &\implies d(x^k) \to 0 \\ &\implies x' \in M, \end{aligned} \tag{2.12}$$

because this is equivalent to $d(x') = 0$.

Let $\bar{k}$ be an index k, after which j_k repeats infinitely. All such indices evidently belong to the set $J(x')$. Hence, by virtue of the note made in the beginning of the proof and relation (2.9) joined with it, we have (for $y = x'$) $\|x^{k+1} - x'\| = \|x^k - x'\| \neq 0$, i.e., the distances from the points x^k and x^{k+1} to x' will be the same for all $k > \bar{k}$, that contradicts to the convergence $x^k \to x'$. The theorem is proved. □

3 Application of Fejér processes to a system of convex inequalities

3.1 Systems of inequalities in $\mathbb{R}^n$

Let

$$f_j(x) \leq 0, \quad j = 1, \dots, m \tag{3.1}$$

be a system of inequalities with convex functions at the left-hand sides (a system of convex inequalities), with the set of solutions satisfying $M \neq \varnothing$. Suppose $\mathbb{R}^n$ to be the domain of definition of the functions $\{f_j(x)\}$. Thus, these functions are differentiable on the whole space $\mathbb{R}^n$ (see (Rockafellar, 1970, p. 23)).

Let $M_j = \{x : f_j(x) \leq 0\}$, $j = 1, \dots, m$, and then $M = \bigcap_{j=1}^m M_j$. We consider the construction (7.3), Chapter I, as basis for the generation of M-Fejér processes for (3.1); in this construction we have $E(x) = \partial d(x)$, i.e.,

$$T(x) = \left\{ x - \lambda \frac{d^+(x)}{\|h\|^2} \cdot h : h \in \partial d(x) \right\}. \tag{3.2}$$

Here, $\lambda \in (0, 2)$, $d(x) = \max_j f_j(x)$. As before, in (3.2) $T(x) = x$ if $d^+(x) = 0$, so we avoid the situation $h = 0$. Actually, for $h = 0 \in \partial d(\bar{x})$

the point $\bar{x}$ is the point of absolute minimum of the convex function $d(x)$. But since $M = \{x : d(x) \leq 0\} \neq \varnothing$, the inequality $d(\bar{x}) \leq 0$ holds, i.e., from $h = 0 \in \partial d(x)$ it follows $d^+(x) = 0$.

Mapping (3.2) is closed (Theorem 7.6, Chapter I), and, therefore,

$$x^{k+1} \in T(x^k) \to x' \in M. \tag{3.3}$$

Put

$$T_j(x) = \left\{ x - \lambda_j \frac{f_j^+(x)}{\|h\|^2} \cdot h : h \in \partial f_j(x) \right\}; \tag{3.4}$$

$$T^{(1)}(x) = \sum_{j=1}^{m} \alpha_j T_j(x), \quad \alpha_j > 0, \ \sum_{j=1}^{m} \alpha_j = 1; \tag{3.5}$$

$$T^{(2)}(x) = T_1 \ldots T_m(x); \tag{3.6}$$

$$T^{(3)}(x) = \left\{ x - \lambda \frac{f^+(x)}{\|h\|^2} \cdot h : h \in \partial f(x) \right\}. \tag{3.7}$$

The choice of the function $f(x)$ in (3.7) obeys the following conditions: $f(x)$ is convex and $\{x : f(x) \leq 0\} = M$ is the set of solutions of system (3.1). In particular, the function $f(x)$ can be of the form $\max_j f_j(x)$ that will correspond to the case of operator (1.5). Relation (3.6) contains the superposition of multi-valued mappings. This follows from the fact that if w_1 and w_2 are multi-valued mappings, then $w_1 w_2(x) = \bigcup_{y \in w_2(x)} w_1(y)$. Constructions (3.5) and (3.6) repeat (1.3) and (1.4), and construction (3.7) corresponds to the construction (1.5) with $f(x) = \max_j f_j(x)$. In (3.7), the function $f(x)$ can be constructed in many ways. In particular,

$$f(x) = \sum_{j=1}^{m} \gamma_j f_j^+(x), \quad \gamma_j > 0, \ j = 1, \ldots, m, \tag{3.8}$$

or, for example,

$$f(x) = \sum_{j=1}^{m} \gamma_j (f_j^+(x))^{\mu}, \tag{3.9}$$

where $\gamma_j > 0$, $\mu > 1$. Note that unlike (3.8), this function is differentiable if the functions $f_j(x)$ are differentiable.

For $f(x)$ given in such a way, the subdifferentials have the form

$$\partial \max_j f_j(x) = \operatorname{conv}\{\partial f_j(x) : j \in J(x)\}, \tag{3.10}$$

$$\partial \sum_{j=1}^m \gamma_j f_j^+(x) = \sum_{j=1}^m \gamma_j \operatorname{conv}\{\partial \max\{f_j^+(x), 0\}\}, \tag{3.11}$$

$$\partial \sum_{j=1}^m \gamma_j (f_j^+(x))^\mu = \sum_{j=1}^m \gamma_j \, \mu \, (f_j^+(x))^{\mu-1} \partial(f_j^+(x), 0) \tag{3.12}$$

(see, for instance, (Dem'yanov and Vasil'ev, 1981, Chapter I, Section 5, and others)). Here, $J(x) = \{j : f(x) = f_j(x)\}$.

3.2 Systems of inequalities in a Hilbert space

As in the previous section, here we shall investigate the system of convex inequalities (3.1), with one difference: now f_j are supposed to be convex subdifferentiable functionals defined on a Hilbert space $\mathcal{H}$.

We start by recalling some definitions.

Definition 3.1. The set $\operatorname{dom} f = \{x \in \mathcal{H} : f(x) < \infty\}$ is called the *effective domain* of the functional f.

Definition 3.2. The functional f is called *proper*, if $\operatorname{dom} f \neq \varnothing$ and $f(x) > -\infty$ for all x.

As it is known (Ioffe and Tikhomirov, 1974), a convex proper functional continuous at a point x_0 is subdifferentiable at this point, and its subdifferential $\partial f(x_0)$ is bounded in a weak topology, hence it is bounded by the norm (Kolmogorov and Fomin, 1974). Particularly, some usual convex function $f : \mathbb{R}^n \to \mathbb{R}^1$ is continuous on $\operatorname{int}\operatorname{dom} f$ (here, $\operatorname{int} Q$ is the set of the interior points of the set Q), thus, this function is subdifferentiable at any point $x \in \operatorname{int}\operatorname{dom} f$, and its subdifferential is both bounded by norm and a bounded mapping from $\mathbb{R}^n$ into $\mathbb{R}^n$, i.e.,

$$\sup_{x\in Q,\, h\in\partial f(x)} \|h\| \le C(Q) < \infty$$

for any bounded set $Q \in \operatorname{int}\operatorname{dom} f$.

We return to the mappings $T^{(i)}$ $(i = 1, 2, 3)$ represented by formulas (3.4)–(3.12). According to the Lemmas 3.13 and 3.14, Chapter I, each of these mapping belongs to the class $\mathcal{P}_M^\nu$, where M is the set of solutions of system

(3.1) and, for this, $M = \mathrm{Fix}(T^{(i)})$. Then from Theorem 3.1, Chapter I, the following corollary follows.

Corollary 3.3. *Let the operator $T : \mathcal{H} \to 2^{\mathcal{H}}$ coincide with one of the mappings $T^{(i)}$ $(i = 1, 2, 3)$, and for the operator T let relation* (3.1)*, Chapter I, be satisfied. Then the process*

$$x^{k+1} \in T(x^k), \quad k = 0, 1, \ldots$$

generates a sequence x^k that converges weakly to some solution $\hat{x}$ of system (3.1)*, and properties 2–4 of Theorem 3.1, Chapter I, are valid.* □

As it was already noted in Section 3, Chapter I (Remark 3.15), the semicontinuity from below of the functionals f_j and boundedness of their subdifferentials guarantee validity of relation (3.1), Chapter I, that is used in the conditions of Corollary 3.3.

4 Systems of convex inclusions

Below we discuss the finite *system of inclusions*

$$x \in M_j, \quad j = 1, 2, \ldots, m, \tag{4.1}$$

in which $\{M_j\}_1^m$ are nonempty convex subsets of $\mathbb{R}^n$. Both the case $M = \bigcap_j M_j \neq \varnothing$ and the case $M = \varnothing$ will be considered.

4.1 General properties and constructions

Denote by $P_j(x)$ the metric projection onto M_j, and $\rho_j(x)$ is the distance from x to M_j, i.e., $\rho_j(x) = \|x - P_j(x)\|$. The following properties of the projection operator $P_j(x)$ are well known (see, for instance, (Berdyshev and Petrak, 1999, Section 3)):

1) the operator $P_j(x)$ is single-valued and continuous;

2) the distance function $\rho_j(x)$ is differentiable, and then,

$$\left.\begin{aligned} \nabla \rho_j(x) &= \frac{x - P_j(x)}{\|x - P_j(x)\|} && \forall\, x \notin M_j, \\ \nabla \rho_j^2(x) &= 2\,(x - P_j(x)) && \forall\, x \in \mathcal{H}; \end{aligned}\right\} \tag{4.2}$$

3) $\|P_j(x) - P_j(y)\| \le \|x - y\|$ $\forall\, x,\ y \in \mathcal{H}$, i.e., the metric projection onto a convex closed set in $\mathcal{H}$ is nonexpansive.

Let

$$T_j(x) = x - \lambda_j[x - P_j(x)], \quad \lambda_j \in (0,2). \tag{4.3}$$

Lemma 4.1. *The mapping* (4.3) *is continuous* M_j*-Fejér, i.e.,*

$$T_j(x) \in \mathcal{F}_{M_j}.$$

Proof. This follows from Lemma 3.13, Chapter I. □

Consider the following mappings,

$$T^{(1)}(x) = \sum_{j=1}^{m} \alpha_j T_j(x), \quad \alpha_j > 0, \ \sum_{j=1}^{m} \alpha_j = 1; \tag{4.4}$$

$$T^{(2)}(x) = T_1 T_2 \dots T_m(x); \tag{4.5}$$

$$T^{(3)}(x) = T_{j_x}(x), \tag{4.6}$$

where $j_x \in J(x) = \{j : \max_j \rho_j(x) = \rho_{j_x}(x)\}$.

Generally speaking, the choice of j_x in (4.6) is not unique, and, as a result, mapping (4.6) is not single-valued. For simplicity we have defined it in the form of an equality, but the following form would be more correct:

$$T^{(3)}(x) = \{T_j(x) : \ j \in J(x)\}. \tag{4.7}$$

In the sequel we consider the form (4.6).

The construction of the mappings (4.4) and (4.5) corresponds completely to the constructions of (3.5) and (3.6). As for mapping (4.6), its correspondence to mapping (3.7) holds, but here, some further explanations are necessary.

The operator $T^{(3)}$ is constructed by mappings (4.3). Using (4.2), the latter ones can be rewritten in the form

$$T_j(x) = x - \lambda_j \frac{\rho_j(x)}{\| \nabla \rho_j(x)\|^2} \nabla \rho_j(x). \tag{4.8}$$

In (4.8) we suppose $x \notin M_j$, otherwise as before, we suppose $T_j(x) = x$. Applying representation (4.8), operator (4.6) can be rewritten in the following equivalent form:

$$T^{(3)}(x) = x - \lambda_{j_x} \frac{\rho_{j_x}(x)}{\|h_x\|^2} h_x, \tag{4.9}$$

where $j_x \in J(x)$, $h_x = \nabla \rho_{j_x}(x) \in \partial d(x)$, $d(x) = \max_j \rho_j(x)$. Thus, it is clear that (4.9) corresponds to (3.7), but in (4.9) we do not take an arbitrary h from $\partial d(x)$, but a special one, namely, $h_x = \nabla \rho_{j_x}(x)$.

Lemma 4.2. *Mapping* (4.6) *(or in another notation,* (4.7)*) is closed in* $\mathbb{R}^n$.

Proof. Let $x_k \to \bar{x}$,

$$y_k = x_k - \lambda_{j_k}[x_k - P_{j_k}(x_k)], \quad y_k \to \bar{y}, \tag{4.10}$$

and then $j_k \in J(x_k)$, i.e.,

$$\underbrace{\max_j \|x_k - P_j(x_k)\|}_{d(x_k)} = \|x_k - P_{j_k}(x_k)\|. \tag{4.11}$$

For closedness of (4.6) it is necessary to show that $\bar{y} \in T^{(3)}(\bar{x})$, i.e.,

$$\bar{y} = \bar{x} - \lambda_{\bar{j}}[\bar{x} - P_{\bar{j}}(x)], \tag{4.12}$$

and then $\bar{j} \in J(\bar{x})$, i.e.,

$$\underbrace{\max_j \|\bar{x} - P_j(\bar{x})\|}_{d(\bar{x})} = \|\bar{x} - P_{\bar{j}}(\bar{x})\|. \tag{4.13}$$

Consider now relations (4.10) and (4.11). Since the index j_k takes a finite number of values, it is possible to select in the sequence $\{k_i\}_i$ a subsequence $\{k_i\}_i$ such that all j_k coincide for $k = k_i$, and this will be denoted by $\bar{j}$. Then relations (4.10) and (4.11) can be rewritten in the form

$$y_{k_i} = x_{k_i} - \lambda_{\bar{j}}[x_{k_i} - P_{\bar{j}}(x_{k_i})],$$
$$\underbrace{\max_j \|x_{k_i} - P_j(x_{k_i})\|}_{d(x_{k_i})} = \|x_{k_i} - P_{\bar{j}}(x_{k_i})\|.$$

Passing to the limit in this relations (with taking, naturally, into account the continuity of the function $d(x)$), we obtain relations (4.12) and (4.13) needed for the completion of the proof. □

4.2 Consistent systems of inclusions

Consistency of system (4.1) means that $M = \bigcap_j M_j \neq \varnothing$. The mappings $T^{(1)}(x)$ and $T^{(2)}(x)$ are continuous and M-Fejér, hence the statement is valid for convergence in $\mathbb{R}^n$ to elements in M and for the sequences generated by these mappings.

Above, it was proved that the mapping $T^{(3)}(x)$ is closed, therefore, according to Theorem 6.10, Chapter I, the statement is valid for convergence of the sequence x^k that is generated by the recurrent relation

$$x^{k+1} \in T^{(3)}(x^k),$$

to some element from the intersection $\bigcap_j M_j$.

In the general case (of an infinite-dimensional) Hilbert space $\mathcal{H}$, the processes under consideration with the iterative operators $T^{(i)}$ $(i = 1, 2, 3)$ converge weakly to some point from the set $M = \bigcap_j M_j$.

Actually, the original operators T_j (formulas (4.3) and (4.8)) are both M_j-Fejér (on the basis of results of Section 3, Chapter I) and, also, belong to the class $\mathcal{P}^{\nu_i}_{M_j}$ of M_j-pseudo-contractive mappings. Then according to Theorem 3.9, Chapter I, the operators $T^{(i)}$ $(i = 1, 2, 3)$ defined by formulas (4.4)–(4.6) belong to the class $\mathcal{P}^{\nu}_{M}$ of M-pseudo-contractive mappings for some $\nu > 0$ (see Theorem 3.9, Chapter I). Since subdifferentials of the functionals $f_j(x)$ are evidently bounded mappings (formula (4.2)), the conditions of Theorem 3.1, Chapter I, are satisfied (see Remark 3.15, Chapter I) that guarantee weak convergence of the iterations and, additionally, the validity of properties 2–4.

To construct a strongly converging sequence of iterations for the approximation of a solution to problem (4.1), it is possible to use the method of correcting multipliers (see formula (4.1), Chapter I). For this we use $\lambda_i = 1$ in formula (4.3). Then the mapping T_j coincides with some metric projection P_j onto the set M_j. Since P_j are nonexpansive operators, the mappings $T^{(i)}$ $(i = 1, 2, 3)$, defined by formulas (4.4)–(4.6) will also belong to the class $\mathcal{K}$ of nonexpansive operators.

Consider some element v_0 which plays the role of the test solution and contains possible information about a solution to be found (if no such information is available, then any element from $\mathcal{H}$ can be taken as v_0; for example, it is possible to take $v_0 = 0$). Denote by $\bar{z}$ the metric projection of the element v_0 onto the set $M = \bigcap_{j=1}^{m} M_j$. Then it is evident that each of the operators $T^{(i)}$ will map the ball $\bar{S}_R(\bar{z})$ into itself (here, $R = \|v_0 - \bar{z}\|$), since $\|T^{(i)}(x) - \bar{z}\| \le \|x - \bar{z}\|$ where $T^{(i)} \in \mathcal{P}^1_M \subset \mathcal{K}_M$ is taken into account.

Thus, we are in the framework of Theorem 4.5, Chapter I, that implies the following corollary.

Corollary 4.3. *Let γ_k be an admissible sequence and $x^0 \in \bar{S}_R(\bar{z})$. Then the sequence x^k generated by the process*

$$x^{k+1} = \gamma_{k+1} T(x^k) + (1 - \gamma_k)\, v_0,$$

where the operator $T = T^{(i)}$ *is defined by one of the formulas* (4.4)–(4.6), *converges strongly to the* v_0*-normal solution of problem* (4.1). □

4.3 Convergent processes for inconsistent systems of inclusions

Generally speaking, the operators $T_j(x) = x - \lambda_j[x - P_j(x)]$ are not nonexpansive, but for $\lambda_j = 1$ (in this case $T_j(x) = P_j(x)$) they become such ones (this was noted in the properties of an projection operator in Section 4.1 of this chapter). Write the operator $T^{(1)}(x)$ for the case $\lambda_j = 1$, $j = 1, \dots, m$, then it takes the form

$$T_0^{(1)}(x) = \sum_{j=1}^{m} \alpha_j P_j(x), \quad \alpha_j > 0, \ \sum_{j=1}^{m} \alpha_j = 1. \tag{4.14}$$

Note that $T_0^{(1)}(x)$ is a nonexpansive operator, i.e.,

$$\|T_0^{(1)}(x) - T_0^{(1)}(y)\| \le \|x - y\| \quad \forall\, x, y \in \mathbb{R}^n. \tag{4.15}$$

Operator (4.14) can be changed to the form

$$T_0^{(1)}(x) = x - \frac{1}{2} \nabla f(x), \tag{4.16}$$

where $f(x) = \sum_{j=1}^{m} \alpha_j \rho_j^2(x)$. From that we obtain

$$\mathrm{Fix}(T_0^{(1)}) = \{x : \ \nabla f(x) = 0\}. \tag{4.17}$$

Denote $\widetilde{M} = \mathrm{Fix}(T_0^{(1)})$. The set $\widetilde{M}$ may be empty. The property $\widetilde{M} \neq \varnothing$ will hold under the so called condition of correctness of the sets system. This notation is introduced next.

Definition 4.4. A system of nonempty, convex, and closed sets $\{M_j\}_1^m$ is called *correct*, if for some $\varepsilon_j \ge 0$ the system of inequalities $\rho_j(x) \le \varepsilon_j$, $j = 1, \dots, m$, has a nonempty and bounded set of solutions. (Recall that $\rho_j(x) = \|x - P_{M_j}(x)\|$.)

The set $M_j^{\varepsilon_j} = \{x : \rho_j(x) \le \varepsilon_j\}$ is also called the ε_j*-expansion* of the set M_j.

Remark to Definition 4.4 (Eremin and Astaf'ev, 1976, Lemma 24.1). Actually, if for some $\varepsilon_j \ge 0$ the set $\bigcap_j M_j^{\varepsilon_j}$ is nonempty and bounded, then for any other $\delta_j \ge 0$ that provide the nonemptyness of the set $\bigcap_j M_j^{\delta_j}$, boundedness holds.

Lemma 4.5. *Let the system* $\{M_j\}_1^m$ *have the property of correctness. Then the operation* inf *in the problem*

$$\tilde{\gamma} = \inf_x \sum_{j=1}^{m} \gamma_j \rho_j^2(x), \quad \gamma_j > 0 \tag{4.18}$$

is achievable, i.e., there exists $\tilde{x} \in \mathbb{R}^n$ *such that*

$$\sum_{j=1}^{m} \gamma_j \rho_j^2(\tilde{x}) = \tilde{\gamma}.$$

Proof. Put $\sum_{j=1}^m \gamma_j \rho_j^2(x) = d(x)$, and let γ be such that $M_d = \{x : d(x) \le \gamma\} \ne \varnothing$. Now prove the boundedness of the set M_d. If $\bar{x} \in M_d$, then from the inequality $d(\bar{x}) \le \gamma$ it follows $\gamma_j \rho_j^2(\bar{x}) \le \gamma$ for all j, or $\rho_j^2(\bar{x}) \le \bar{\gamma}_j$ $(= \gamma / \gamma_j)$, $j = 1, \dots, m$. From the condition of correctness for $\{M_j\}_1^m$, the boundedness (as it was already noted) of solutions follows for the latter system of inequalities. Since $\bar{x}$ is an arbitrary element from the set M_d, the boundedness of this set follows from this.

Now take some monotone sequence of numbers $\delta_k \to 0$, $\delta_k > 0$ and consider the sequence of sets $N_k = \{x : d(x) \le \tilde{\gamma} + \delta_k\}$, $k = 1, 2, \dots$ that have the property $N_k \supset N_{k+1}$ $\forall k$. By virtue of convexity, closedness, and boundedness of these sets, the intersection $\bigcap_k N_k = \widetilde{N}$ is not empty. Taking a convergent subsequence x_k, $x_k \in N_k$, we shall have $x_k \to \tilde{x} \in \widetilde{N}$ and, for this, $d(\tilde{x}) \le \tilde{\gamma}$; and $d(\tilde{x}) = \tilde{\gamma}$ follows evidently, which has to be proved. □

Now we can formulate the following theorem.

Theorem 4.6. *Let* $T_0^{(1)}(x)$ *be the operator given accordingly to* (4.14)*; and* $f(x) = \sum_{j=1}^m \alpha_j \|x - P_j(x)\|^2$, $\widetilde{M} = \{x : \nabla f(x) = 0\}$. *If for* $\{M_j\}_1^m$ *the condition of correctness (see Definition 4.4) is satisfied, then*

1) $\widetilde{M} = \mathrm{Fix}(T_0^{(1)}) \ne \varnothing$;

2) $T_1(x) = (1-\alpha)T_0^{(1)}(x) + \alpha x$, $T_1 \in \mathcal{F}_{\widetilde{M}}$, $\alpha \in (0,1)$;

3) $T_1^k(x_0) \to \tilde{x} \in \widetilde{M}$, $k \to \infty$.

Proof. The fact $\widetilde{M} = \mathrm{Fix}(T_0^{(1)})$ follows from (4.16). The nonemptyness of the set $\widetilde{M}$ is, in essence, the contents of Lemma 4.5, since the equality $\nabla f(x) = 0$ for a convex differentiable function $f(x)$ is a necessary and sufficient condition for the point x to be an absolute minimizer of the function $f(x)$ in $\mathbb{R}^n$.

Property 2 is valid, since the projection operators $P_j(x)$ are nonexpansive, and, thus, $T_0^{(1)}(x)$ is also nonexpansive. Therefore, $T_0^{(1)}(x)$ is weakly $\widetilde{M}$-Fejér and, so, $T_1 \in \mathcal{F}_{\widetilde{M}}$.

As for property 3, the mentioned convergence holds by virtue of continuity of $T_1(x)$ and since it is $\widetilde{M}$-Fejér. □

4.4 Cycles of immobility for inconsistent systems of convex inequalities

Let $T_j : \mathbb{R}^n \to M_j$, $j = 1, \dots, m$. Such mappings can be the operators $P_j(x)$ of projection of x onto M_j, where M_j are convex closed sets. Consider

$$T(x) = T_1 T_2 \dots T_m(x).$$

Definition 4.7. An ordered system of vectors $\{\bar{x}_1, \bar{x}_2, \dots, \bar{x}_m\}$, $\bar{x}_j \in M_j$ is called a *cycle of immobility* $T(x)$ in application to system (4.1) if the following relations hold:

$$\left.\begin{array}{lcl} T_m(\bar{x}_1) & = & \bar{x}_m, \\ T_{m-1}(\bar{x}_m) & = & \bar{x}_{m-1}, \\ & \vdots & \\ T_2(\bar{x}_3) & = & \bar{x}_2, \\ T_1(\bar{x}_2) & = & \bar{x}_1. \end{array}\right\} \tag{4.19}$$

Relations (4.19) can be represented by the following scheme:

$$\underbrace{T_1\ \overbrace{T_2\ \dots\ \overbrace{T_{m-1}\ \underbrace{T_m\,(\bar{x}_1)}_{\bar{x}_m}}^{\bar{x}_{m-1}}}^{\bar{x}_2}}_{\bar{x}_1} = \bar{x}_1$$

Scheme 4.1

Note that if $\bigcap_j M_j \neq \varnothing$, then $\bar{x}_1 = \bar{x}_2 = \dots = \bar{x}_m \in \bigcap_j M_j$, and the general value for $\bar{x}_i$, say, $\bar{x}$, is the point of the mentioned intersection.

The notion of the cycle of immobility can be regarded as an extension of a solution of the inclusions system (4.1).

Theorem 4.8. *Let $\{T_j\}_1^m$ be nonexpansive operators, and let one of the sets $\{M_j\}_1^m$, say, M_1, be bounded. Then there exists at least one cycle of immobility.*

Proof. Actually, it is necessary to prove existence of some fixed point for $T(x) = T_1 T_2 \dots T_m(x)$. Since the operator $T(x) = T_1 T_2 \dots T_m(x)$, firstly, is continuous and, secondly, transfers the convex, bounded, and closed set M_1 into itself, by the Schauder–Tikhonov theorem (see, for instance, (Edwards, 1969)) the operator $T(x)$ has at least one, say, $\bar{x}_1$, fixed point lying in M_1. If the element $\bar{x}_1$ is added to the elements $\bar{x}_m, \dots, \bar{x}_2$ obtained by formulas (4.19), then, as a result, we obtain the set $\{\bar{x}_1, \dots, \bar{x}_m\}$ which satisfies the definition of an immobility cycle. □

Corollary 4.9. *If $T_j(x) = P_j(x)$, and if one of the sets $\{M_j\}_1^m$, say M_1, is bounded, then the statement of Theorem 4.8 is valid.*

Proof. Follows from the nonexpansiveness of the operators $P_j(x)$ and, hence, from the nonexpansiveness of $T(x) = P_1 P_2 \dots P_m(x)$. □

Let $\{T_j(x)\}_1^m$ and $\{M_j\}_1^m$ satisfy the conditions of Theorem 4.8, $T(x) = T_1 T_2 \dots T_m(x)$. Let $\widetilde{M} = \text{Fix}(T)$. By Theorem 4.8 we have $\widetilde{M} \neq \varnothing$. By virtue of the property of nonexpansiveness of the operator $T(x)$ we have:

$$T(y) = y, \quad \|T(x) - y\| \leq \|x - y\| \quad \forall\, y \in \widetilde{M}, \ \forall\, x \notin \widetilde{M},$$

i.e., $T(x)$ is weakly $\widetilde{M}$-Fejér, therefore,

$$T_\alpha(x) = (1 - \alpha)T(x) + \alpha x, \quad \alpha \in (0, 1) \tag{4.20}$$

is $\widetilde{M}$-Fejér.

Thus, the following theorem is valid.

Theorem 4.10. *The operator* (4.20) *is a continuous $\widetilde{M}$-Fejér mapping and, from this,*

$$T_\alpha^k(x_0) \xrightarrow{(k)} \bar{x}_1 \in \widetilde{M}.$$ □

Remark to Theorem 4.10. Since $\text{Fix}(T) = \text{Fix}(T_\alpha)$, we have

$$T_\alpha(\bar{x}_1) = \overline{x}_1 = T(\overline{x}_1).$$

Other elements of the immobility cycle are reconstructed according to formulas (4.19) or according to Scheme 4.1.

5 On the rate of convergence of Fejér processes

The Fejér processes under consideration generated by the mappings $T \in \mathcal{F}_M$ provide a geometrical progression convergence in the following sense:

$$|x^{k+1} - M| \leq \Theta\,|x^k - M|, \tag{5.1}$$

where x^k is the sequence given by the relation $x^{k+1} \in T(x^k)$ (we take the case of a pointwise-multiple mapping T), and where $\Theta \in (0,1)$, $|x - M| = \inf_{y \in M} \|x - y\|$.

For each of the mappings (3.5), (3.6), and (3.7), it is possible to find the rate of the geometrical progression convergence for the sequences generated by these mappings (under these or that nonrigid constraints). For an illustration consider case (3.7), i.e.,

$$T(x) = \left\{ x - \lambda \frac{f^+(x)}{\|h\|^2} h : h \in \partial f(x) \right\}, \tag{5.2}$$

$\lambda \in (0,2)$, $h \in \partial f(x)$.

We are interested in the case $f(x) = \max_j f_j(x)$, where $\{f_j(x)\}_1^m$ are from system (3.1). In (5.2) we put, as before, $T(x) = x$ if $h = 0$. We shall assume that $M = \{x : f(x) \leq 0\} \neq \varnothing$.

For the analysis of the methods for solving systems of linear and convex inequalities, one often uses the so called *Slater condition* and the Hoffmann lemma (Hoffmann, 1952). For the convex inequality $f(x) \leq 0$ the first condition means that there exists a p with $f(p) < 0$. For the system of convex inequalities $f_j(x) \leq 0$, $j = 1, \ldots, m$, this means: $f(p) = \max_j f_j(p) < 0$ for some p.

The statement of the *Hoffmann lemma* in our situation provides the existence of a constant $C > 0$ such that

$$|x - M| \leq C\, f^+(x), \quad x \in S, \tag{5.3}$$

where the set S is chosen by this or that way in dependence on the situation. If for the convex inequality $f(x) \leq 0$ the Slater condition is satisfied and S is a convex bounded set, then relation (5.3) is valid.

In the case of the system of linear inequalities $Ax \leq b$, the Hoffmann lemma can also be written in the following form

$$|x - M| \leq C\, \|(Ax - b)^+\|,$$

where $M = \{x : Ax \leq b\} \neq \varnothing$. Here, the validity of the Slater condition is not obligatory. Note that questions on majorizing the function $f(x)$ in the form (5.3) were considered in the work (Eremin, 1968). There, the estimates of the convergence rate were also obtained for some Fejér sequences generated by the operators $T \in \mathcal{F}_M$.

By means of mapping (5.2), define the process

$$x^{k+1} = x^k - \lambda \frac{f(x^k)}{\|h^k\|^2} h^k, \tag{5.4}$$

where $\lambda \in (0, 2)$, $h^k \in \partial f(x^k)$. We shall assume that $\{x^k\} \cap M = \varnothing$. In this case $0 \notin \{h^k\}$. Actually, if $h^{\bar{k}} = 0$, then from the inequality $(h^{\bar{k}}, x - x^{\bar{k}}) \le f(x) - f(x^{\bar{k}})$ that holds for any convex function and arbitrary x, it would follow $f(x^{\bar{k}}) \le f(x)\ \forall\, x \in \mathbb{R}^n$, and then $x^{\bar{k}}$ would be an absolute minimizer of the function $f(x)$. However, this contradicts to the fact that $f(x^{\bar{k}}) > 0$ (since $x^{\bar{k}} \notin M$) and $M = \{x : f(x) \le 0\} \ne \varnothing$.

Theorem 5.1. *Let the condition* (5.3) *be satisfied and the set* $\operatorname{conv}\{x^k\}$ *be taken as the set* S. *Then for the sequence* x^k *generated by relation* (5.4), *estimate* (5.1) *holds for*

$$\Theta = [1 - \lambda\,(2 - \lambda) C^{-2} \gamma_0^{-2}]^{1/2}, \tag{5.5}$$

where $\gamma_0 = \sup_k \|h^k\|$.

Proof. 1. Since the sequence x^k is bounded, it follows by Lemma 7.3, Chapter I,

$$\sup_k \|h^k\| = \gamma_0 < +\infty. \tag{5.6}$$

From condition (5.3) it follows

$$\frac{f^2(x^k)}{|x^k - M|^2} \ge C^{-2} \quad \forall\, k. \tag{5.7}$$

2. We next establish estimate (5.1). Let $\bar{x}^k$ be the projection of x^k onto M, i.e., we have $|x^k - M| = \|x^k - \bar{x}^k\|$. There holds

$$\begin{aligned}
|x^{k+1} - M|^2 &\le \|x^{k+1} - \bar{x}^k\|^2 \\
&= \|x^k - \bar{x}^k\|^2 + 2\lambda \frac{f(x^k)}{\|h^k\|^2} (h^k, \bar{x}^k - x^k) + \lambda^2 \frac{f^2(x^k)}{\|h^k\|^2} \\
&\le |x^k - M|^2 + 2\lambda \frac{f(x^k)}{\|h^k\|^2} [f(\bar{x}^k) - f(x^k)] + \lambda^2 \frac{f^2(x^k)}{\|h^k\|^2}
\end{aligned}$$

$$= |x^k - M|^2 - \lambda(2-\lambda)\frac{f^2(x^k)}{\|h^k\|^2}$$

$$= \left[1 - \frac{\lambda\,(2-\lambda)}{\|h^k\|^2}\,\frac{f^2(x^k)}{|x^k - M|^2}\right]|x^k - M|^2$$

$$\overset{(5.6)\,\&\,(5.7)}{\leq} \left[1 - \frac{\lambda\,(2-\lambda)}{\gamma_0^2}\,C^{-2}\right]|x^k - M|^2 = \Theta^2|x^k - M|^2.$$

Thus, estimate (5.1) is established, with the value Θ given according to (5.5).□

6 Fejér methods and nonsmooth optimization

Under the term "*nonsmooth optimization*" one understands the mathematical techniques and methods for solving problems of mathematical programming without properties of smoothness (differentiability) of the functions that form these problems. The following components belong to the range of such techniques: convex analysis (Rockafellar, 1970), the subdifferentiability (and its properties), the techniques for reducing problems with constraints to nonsmooth problems without constraints, and others (Rockafellar, 1970; Shor, 1979; Polyak, 1969; Eremin, 1969; Dem'yanov and Vasi'ev, 1981; Clark, 1988; and others).

The methods of nonsmooth optimization are sufficiently well elaborated for problems of convex programming, the latter by using the techniques of the exact penalty functions are reduced to simple problems of the form

$$\min_x \Phi(x), \tag{6.1}$$

where $\Phi(x)$ is a non-differentiable convex function defined on $\mathbb{R}^n$, and by virtue of this, the function at each point x has the subdifferential $\partial f(x) \neq \varnothing$. Problem (6.1) can be a maximum problem, but the function $\Phi(x)$ must be concave then.

The Fejér technology is well adjusted to solving nonsmooth systems of convex inequalities and problems of convex programming (Eremin, 1969).

Consider the problem of convex programming

$$\max\{f(x) : \ f_j(x) \leq 0, \ j = 1, \dots, m\} \tag{6.2}$$

and, separately, the system of convex inequalities

$$f_j(x) \leq 0, \quad j = 1, \dots, m. \tag{6.3}$$

If we say that (6.2) is a problem of convex programming, then this means that the functions $\{-f(x), f_1(x), \dots, f_m(x)\}$ are convex.

6.1 Problem of the saddle point of the Lagrange function

In this section the linear space $\mathcal{X}$ of the variable $x \in \mathcal{X}$ is arbitrary.

In correspondence to problem (6.2), we consider the following *Lagrange function*

$$F(x,u) = f(x) - \sum_{j=1}^{m} u_j f_j(x), \tag{6.4}$$

where $u_j \geq 0$, $j = 1, \dots, m$, are the *Lagrange multipliers*. Immediately note that if we consider in (6.2) min instead of max then the Lagrange function will have the form

$$F(x,u) = f(x) + \sum_{j=1}^{m} u_j f_j(x).$$

Definition 6.1. The point $[\bar{x}, \bar{u}] \in \mathcal{X} \times \mathbb{R}^n$, $\bar{u} \geq 0$ is called a *saddle point* for the Lagrange function (6.4) if the following inequalities are satisfied:

$$F(x,\bar{u}) \underset{\forall x}{\leq} F(\bar{x},\bar{u}) \underset{\forall u \geq 0}{\leq} F(\bar{x},u). \tag{6.5}$$

The following well-known fact is valid (see, for example, (Arrow et al., 1958)).

Theorem 6.2. *Without any assumption on the linear space $\mathcal{X}$ and the functions $\{f(x), f_1(x), \dots, f_m(x)\}$, the following statement is valid: if $[\bar{x}, \bar{u}]$ is a saddle point of the function $F(x,u)$, then $\bar{x} \in \arg(6.2)$ and $\bar{u}_j f_j(\bar{x}) = 0$, $j = 1, 2, \dots, m$.*

Proof. The proof of the theorem is very simple and will be given for completeness of description.

1. Firstly, prove the admissibility of $\bar{x}$ with respect to the constraints in problem (6.2). Consider in more detail relation (6.5):

$$\begin{aligned} f(x) - \sum_{j=1}^{m} \bar{u}_j f_j(x) &\underset{\forall x}{\leq} f(\bar{x}) - \sum_{j=1}^{m} \bar{u}_j f_j(\bar{x}) \\ &\underset{\forall u \geq 0}{\leq} f(\bar{x}) - \sum_{j=1}^{m} u_j f_j(\bar{x}). \end{aligned} \tag{6.6}$$

If for some j, say, for $j = 1$, the inequality $f_1(\bar{x}) > 0$ holds, then taking $u_2 = u_3 = \cdots = u_m = 0$ in the right inequality of (6.6), we obtain

$\sum_{j=1}^{m} \bar{u}_j f_j(\bar{x}) \geq u_1 f_1(\bar{x})$, which gives a contradiction for sufficiently large $u_1 > 0$.

2. We now prove that the following relations hold

$$\bar{u}_j f_j(\bar{x}) = 0, \quad j = 1, \ldots, m. \tag{6.7}$$

Assume the contrary, i.e., that there exists j, say, $j = 1$, such that $\bar{u}_1 f_1(\bar{x}) < 0$. Then the right inequality in (6.6) gives for $u = 0$:

$$0 > \sum_{j=1}^{m} \bar{u}_j f_j(\bar{x}) \geq 0,$$

that gives the contradiction.

3. At last, prove that $\bar{x} \in \arg(6.2)$. The left inequality in (6.6), by taking into account that relations (6.7) are already proved, can be rewritten in the form $f(x) \leq f(\bar{x}) + \sum_{j=1}^{m} \bar{u}_j f_j(x)$, and from this we obtain $f(x) \leq f(\bar{x})$ for any $x \in M = \{x : f_j(x) \leq 0,\ j = 1, 2, \ldots, m\}$, which means that $\bar{x}$ is optimal for problem (6.2). The theorem is proved. □

Formulation of the inverse theorem requires the introduction of some restrictions. For such a theorem one usually uses the *Kuhn–Tucker theorem* (see, for example, (Arrow et al., 1958)).

Theorem 6.3. *Let $\mathcal{X} = \mathbb{R}^n$, and let $\{-f(x), f_1(x), \ldots, f_m(x)\}$ be convex functions, and let the corresponding system of constraints* (6.3) *of problem* (6.2) *satisfy the condition of regularity, say, in the form of a weakened Slater condition: there exists a solution $\bar{x}$ of system* (6.3) *such that $f_j(\bar{x}) < 0$ for nonlinear $f_j(x)$ (from all $f_j(x)$, $j = 1, 2, \ldots, m$).*

Under the formulated conditions, the following statement is valid: if $\bar{x} \in$ arg (6.2), *then there exists a nonnegative vector $\bar{u} \in \mathbb{R}^m$ such that the vector $[\bar{x}, \bar{u}]$ is a saddle point for the Lagrange function* (6.4).

Proof. The proof of this theorem is based on the theorem about separability of nonintersecting convex sets. □

Remark 6.4. It is easily seen that relations (6.5) are equivalent to the equality

$$\max_{x} \min_{u \geq 0} F(x, u) = \min_{u \geq 0} \max_{x} F(x, u).$$

Remark 6.5 (see, for instance, (Eremin, 1999)). If (6.2) is the problem of linear programming

$$L: \max\{(c,x) : Ax \leq b\}$$

and

$$F_0(x,u) = (c,x) - (u, Ax - b)$$

is its Lagrange function, then the conditions

1^0. $[\bar{x}, \bar{u}]$ is the saddle point for the function $F_0(x,u)$.

2^0. $[\bar{x}, \bar{u}] \in \arg L \times \arg L^*$

are equivalent; here,

$$L^*: \min\{(b,u) : A^\top u = c,\ u \geq 0\}.$$

6.2 Method of the exact penalty functions

A solvable problem of convex programming in the form (6.2) over an arbitrary linear real space with the condition of regularity can be reduced to an equivalent nonsmooth problem of searching for the maximum of the concave function

$$\max_x \Phi_R(x), \tag{6.8}$$

where, in particular, $\Phi_R(x)$ can be of the form

$$\Phi_R(x) = f_0(x) - \sum_{j=1}^{m} R_j f_j^+(x) \tag{6.9}$$

or

$$\Phi_R(x) = f_0(x) - R_0 \max_j f_j^+(x).$$

Here, $\{R_k > 0\}_{k=0}^m$ are positive parameters, for which an appropriate choice provides the equivalent reducibility of a problem with constraints to a problem without constraints. The technology of such a reduction is called the *method of the penalty functions*. There exist more general constructions for the function $\Phi_R(x)$, but we shall consider only one of them in more detail, namely (6.9). The following theorem about the *exact penalty functions* is valid, see (Eremin, 1967 and 1999).

Theorem 6.6. *Let $[\bar{x}, \bar{u}]$ be a saddle point of the Lagrange function $F(x,u)$, i.e., function* (6.4). *If $R = [R_1, \ldots, R_m] \geq \bar{u}$, then*

$$\operatorname{opt}\sup_x \Phi_R(x) = \operatorname{opt}(6.2); \tag{6.10}$$

if $R > \bar{u}$, *then*

$$\arg\sup_x \Phi_R(x) = \arg(6.2). \tag{6.11}$$

Here, opt (6.2) *is the optimal value of problem* (6.2), *and* arg (6.2) *is its optimal set.*

Proof. 1. Firstly, we prove equality (6.10). According to Theorem 6.2 we have $\bar{x} \in \arg(6.2)$, i.e., $\bar{x}$ is the optimal vector for problem (6.2), and this in particular gives $\sum_{j=1}^m R_j f_j^+(\bar{x}) = 0$; from this $\Phi_R(\bar{x}) = f(\bar{x}) = \operatorname{opt}(6.2)$, hence,

$$\sup_x \Phi_R(x) \geq \operatorname{opt}(6.2). \tag{6.12}$$

On the other side, for any $x \in \mathcal{X}$ we have:

$$\left.\begin{aligned} \Phi_R(x) &= f(x) - \sum R_j f_j^+(x) \\ &\overset{(6.6),(6.7)}{\leq} f(\bar{x}) + \sum_{j=1}^m \bar{u}_j f_j(x) - \sum_{j=1}^m R_j f_j^+(x) \\ &= \operatorname{opt}(6.2) - \sum_{j=1}^m (R_j - \bar{u}_j) f_j^+(x) \\ &\leq \operatorname{opt}(6.2) \end{aligned}\right\} \tag{6.13}$$

(in the latter estimate the condition $R \geq \bar{u}$ is taken into account). Since the inequality $\Phi_R(x) \leq \operatorname{opt}(6.2)$ is valid for any $x \in \mathcal{X}$, the following inequality is also valid,

$$\sup_x \Phi_R(x) \leq \operatorname{opt}(6.2).$$

From this, by taking into account (6.12), relation (6.10) follows.

2. Now prove relation (6.11). Let $\tilde{x} \in \arg(6.2)$. But then, on one hand, $\Phi_R(\tilde{x}) = f(\tilde{x}) - \sum_{j=1}^m R_j f_j^+(\tilde{x}) = f(\tilde{x}) = \operatorname{opt}(6.2) = \operatorname{opt}\sup \Phi_R(x)$. On the other hand, according to (6.13) we have

$$\Phi_R(\tilde{x}) \leq \operatorname{opt}(6.2),$$

from this

$$\Phi_R(\tilde{x}) = \sup_x \Phi_R(x),$$

i.e.,

$$\tilde{x} \in \arg\max_x \Phi_R(x).$$

Let now $\tilde{x} \in \arg\max_x \Phi_R(x)$. For the inclusion $\tilde{x} \in \arg(6.2)$ it is necessary to prove only admissibility of the vector $\tilde{x}$ with respect to the constraints of problem (6.2), i.e., $f_j(\tilde{x}) \leq 0$, $j = 1, \dots, m$. But according to (6.13) and the proved relation (6.10), we have:

$$\operatorname{opt} \max_x \Phi_R(x) = \operatorname{opt}(6.2) - \sum_{j=1}^{m} (R_j - \bar{u}_j) f_j^+(\tilde{x}),$$

that gives $\sum_{j=1}^{m} (R_j - \bar{u}_j) f_j^+(\tilde{x}) = 0$; from this and by virtue of the condition $R > \bar{u}$, we obtain the needed inequalities $f_j(\tilde{x}) \leq 0$, $j = 1, \dots, m$.

The theorem is completely proved. □

By virtue of the Kuhn–Tucker theorem (Theorem 6.3 and Theorem 6.6), the following theorem is valid.

Theorem 6.7. *If the problem of convex programming* (6.2) *is solvable and its constraints satisfy the weakened Slater condition, then for the penalty function* $\Phi_R(x)$, *relations* (6.10) *and* (6.11) *are fulfilled, and the vector* $\bar{u} \geq 0$ *there has the meaning of the second component of the saddle point* $[\bar{x}, \bar{u}]$ *of the Lagrange function* $F(x, u)$, *where the existence is provided by the Kuhn–Tucker theorem.*

□

6.3 Fejér methods for the application to systems of convex inequalities and convex programming problems without assumptions of smoothness

Consider the problem of type (6.1), namely,

$$\min_x f(x). \tag{6.14}$$

Here, the choice of the function $f(x)$ will be related either to the problem of searching a solution of the system of convex inequalities

$$f_j(x) \leq 0, \quad j = 1, \dots, m, \tag{6.15}$$

or to the problem of convex programming

$$\min\{f(x) : f_j(x) \leq 0, \ j = 1, \dots, m\}. \tag{6.16}$$

Here, for convenience the problem (6.2) is written in the form of a problem on minimum search that requires convexity of the function $f(x)$.

We shall assume that system (6.15) and problem (6.16) are solvable and, in addition, both system (6.15) and the system of constraints in (6.16) satisfy the weakened Slater condition that by the Kuhn–Tucker theorem provides existence of the saddle point $[\bar{x}, \bar{u}]$, $\bar{u} \geq 0$ for the Lagrange function $F(x,u) = f(x) + \sum_{j=1}^{m} u_j f_j(x)$, $u = [u_1, \dots, u_m] \geq 0$. In application to problem (6.16), the definition of the saddle point is given (unlike of (6.5)) through satisfaction of the inequalities

$$F(x,\bar{u}) \underset{\forall x}{\geq} F(\bar{x},\bar{u}) \underset{\forall u \geq 0}{\geq} F(\bar{x},u).$$

Consider a reduction of the problem of solvong the system of convex inequalities, say, system (6.15), to problem (6.14). This is carried out trivially; it is sufficient to consider, for example,

$$f(x) = \sum_{j=1}^{m} R_j f_j^+(x), \tag{6.17}$$

$R_j > 0$, $j = 1, \dots, m$.

The following properties are evident:

1) the function $f(x)$ is convex;

2) the set of solutions of system (6.15) coincides with $\arg\min_x f(x)$;

3) $\arg\min_x f(x) = \{x : f(x) \leq 0\}$.

Here, a particular situation arises when for the solution of the inequality

$$f(x) \leq 0, \tag{6.18}$$

to which system (6.15) was reduced, the following iterative process (see Section 7, Chapter I) is applied:

$$\begin{gathered} x^{k+1} \in T(x^k), \quad k = 0, 1, 2, \dots, \\ T(x) = \left\{ x - \lambda \frac{f^+(x)}{\|h\|^2} h : h \in \partial f(x) \right\}, \quad 0 < \lambda < 2. \end{gathered} \tag{6.19}$$

The following theorem is valid.

Theorem 6.8. *Under the conditions mentioned above, process* (6.19) *converges to some solution x' of inequality* (6.18), *and then x' is a solution of system* (6.15). □

In the situation of solvability of system (6.15), the minimum of the function $f(x)$ is equal to zero, and we have already used this fact. It is possible to consider a more general situation when $f(x)$ is an arbitrary convex function with a finite and achievable minimum $\tilde{f} = \operatorname{opt}_x f(x)$.

Then the problem of determining $\tilde{x} \in \arg\min_x f(x)$ can be formulated as the problem of solving the convex inequality

$$f(x) \le \tilde{f},$$

that is reduced to process (6.19) with $T = T'$ where

$$T'(x) = \left\{ x - \lambda \frac{f(x) - \tilde{f}}{\|h\|^2} \cdot h \,:\, h \in \partial f(x) \right\}, \quad 0 < \lambda < 2.$$

Now consider the reduction of problem (6.2) (in formulation for the minimum search) to the problem

$$\min_{x \in \mathbb{R}^n} \Phi_R(x) \tag{6.20}$$

for $\Phi_R(x) = f_0(x) + \sum_{j=1}^m R_j f_j^+(x)$, $R_j > 0$, $j = 1, \dots, m$. If opt (6.20) is known, then the determination of $\tilde{x} \in \arg(6.20)$ could be reduced to the situation just discussed. Otherwise, we have the general case of searching the minimum of a nonsmooth convex function.

6.4 The basic process

For the beginning, consider one meaningful example.

Example (Eremin, 1962). Let the following inconsistent system of linear inequalities be given,

$$l_j(x) = (a_j, x) - b_j \le 0, \quad j = 1, \dots, m. \tag{6.21}$$

The number $E = \min_x \max_j l_j(x) > 0$ is called the *Chebyshev deviation* of system (6.21) (or its *defect*). Denote $d(x) = \max_j l_j(x)$. The function $d(x)$ is convex and piecewise-linear. The point $\bar{x} \in \arg\min_x d(x)$ is called the *point of the Chebyshev deviation* of system (6.21). To find this point, the following iterative process was suggested in the mentioned work:

$$x^{k+1} = x^k - \lambda_k \, d(x^k) \, a_{j_k}, \quad k = 0, 1, \dots, \tag{6.22}$$

where

$$\lambda_k \geq 0, \quad \lambda_k \to 0, \quad \sum_{k=0}^{\infty} \lambda_k = +\infty, \tag{6.23}$$

$$j_k \in J(x^k) = \{j : l_{j_k}(x^k) = d(x^k)\}.$$

Under the made assumptions, the sequence x^k generated by process (6.22) converges to the point of the Chebyshev deviation of system (5.21), and the sequence $d(x^k)$ converges to the Chebyshev deviation E.

It is necessary to pay attention to the fact that the element a_{j_k} in (6.22) is one of the elements of the subdifferential $\partial d(x)$ of the function $d(x)$ at the point x^k. But the subdifferential itself is described by the formula

$$\partial d(x) = \Big\{h : h = \sum_{j \in J(x)} \alpha_j a_j, \ \alpha_j \geq 0, \ \sum_{j=1}^{m} \alpha_j = 1\Big\}.$$

The process of type (6.22) in the general case (i.e., in the case of problem (6.1) under the assumption of convexity of the function $\Phi(x)$) can be written in the following form (Shor, 1979),

$$x^{k+1} = x^k - \lambda_k \frac{h^k}{\|h^k\|}, \qquad k = 0, 1, \ldots, \tag{6.24}$$

where the sequence λ_k satisfies conditions (6.23), $h^k \in \partial\Phi(x^k)$. Process (6.24) is called *basic*. The following theorem is valid (Shor, 1979, Theorem 2.2).

Theorem 6.10. *Let the set*

$$\widetilde{M} = \arg\min_x \Phi(x) \neq \varnothing$$

be bounded, and let for the sequence λ_k conditions (6.23) *be fulfilled. Then:*

1) *if in the course of iteration for the sequence x^k according to* (6.22) *we have $h^{\bar{k}} = 0$ for some $\bar{k}$, then $x^{\bar{k}} \in \widetilde{M}$, and the process is finished;*
2) *if $h^k \neq 0 \ \forall k$, then $\rho(x^k, \widetilde{M}) \to 0$; here, $\rho(x^k, \widetilde{M})$ is the distance function from x^k to $\widetilde{M}$.* □

Remark 6.11. If as function $\Phi(x)$ the function $\Phi_R(x)$ is taken (the latter is given according to (6.20) under condition $R > 0$ that provides validity of

Theorem 6.7), then in the process (6.24) applied to the function $\Phi_R(x)$ the subgradients h^k are calculated by the formula

$$\left.\begin{aligned} h^k &= h_0^k + \sum_{j=1}^{m} R_j h_j^k, \quad h_0^k \in \partial f(x^k), \\ h_j^k &\in \partial f_j(x^k), \quad j = 1, 2, \dots, m. \end{aligned}\right\} \tag{6.25}$$

Under differentiability of all functions $\{f_s(x)\}_{s=0}^m$, relation (6.25) takes the form

$$h^k = \nabla f_0(x^k) + \sum_{j=1}^{m} R_j \nabla f_j(x).$$

Remark 6.12. If we replace the problem $\min_x \Phi(x)$ by a more general one,

$$\min_{x \in S} \Phi(x), \tag{6.26}$$

where S is a convex closed set, then Theorem 6.10 remains valid if the process (6.24) is replaced by the process

$$x^{k+1} = P_S \left[x^k - \lambda_k \frac{h^k}{\|h^k\|} \right].$$

Here $P_S(\cdot)$ is the metric projection onto the set S. In the case of differentiability of $\Phi(x)$, this fact was established in the work (Eremin, 1966) in application to the process

$$x^{k+1} = P_S\Big(x^k - \lambda_k \frac{\Phi(x^k)}{\|\nabla\Phi(x^k)\|^2} \cdot \nabla\Phi(x^k) \Big)$$

under the additional condition $\Phi(x^k) \geq 0 \ \forall\, x \in S$.

7 Aspects of stability of Fejér processes

Let $T \in \mathcal{F}_M$, $M \subset \mathbb{R}^n$. If T is continuous, then according to Theorem 6.2, Chapter I,

$$\{x^{t+1} = T(x^t)\}_{t=0}^{+\infty} \to x' \in M. \tag{7.1}$$

By this, the mapping $x_0 \xrightarrow{\tau} x'$ is defined which puts the limit of the sequence (7.1) in correspondence to the original element x^0.

Definition 7.1. The discrete process (7.1) is called *stable* at the point x^0 if the mapping $\tau(x)$ is continuous in some neighborhood of the point x^0.

Theorem 7.2. *If the mapping $T \in \mathcal{F}_M$ is nonexpansive, then the process* (7.1) *is stable for any original point $x^0 \in \mathbb{R}^n$.*

Proof. Let us prove the nonexpansive property of the mapping $\tau(x)$. Let x^0 and y^0 be arbitrary points from $\mathbb{R}^n$. Since the operator T is nonexpansive (hence, continuous), we have $x^t \to x' \in M$, $y^t \to y' \in M$ (here, $x^t = T^t(x^0)$, $y^t = T^t(y^0)$) and

$$\begin{aligned}\|T^t(x^0) - T^t(y^0)\| &\le \|T^{t-1}(x^0) - T^{t-1}(y^0)\| \\ &\le \cdots \\ &\le \|x^0 - y^0\|.\end{aligned}$$

From this

$$\begin{aligned}\|x' - y'\| &\le \|T^t(x^0) - x'\| + \|T^t(y^0) - y'\| + \|T^t(x^0) - T^t(y^0)\| \\ &\le \|T^t(x^0) - x'\| + \|T^t(y^0) - y'\| + \|x^0 - y^0\|.\end{aligned} \tag{7.2}$$

With thus have

$$\lim_{t\to+\infty} T^t(x^0) = x', \qquad \lim_{t\to+\infty} T^t(y^0) = y',$$

and then passing to the limit in (7.2), we obtain

$$\|x' - y'\| = \|\tau(x^0) - \tau(y^0)\| \le \|x^0 - y^0\|,$$

that gives the continuity of $\tau(x)$ at any point $x^0 \in \mathbb{R}^n$. □

Chapter IV
Some topics of Fejér mappings and processes

1 Decomposition and parallelization of Fejér processes

The structure of the Fejér mappings and processes generated by them gives a wide range of possibilities for their parallelization. This is determined, firstly, by the possibility to construct these iterative mappings by the partial mappings related to the subsystems of the original system, and partition of the system into the subsystems can be arbitrary. Second, the structure of the matrix of constraints (block-wise, diagonal, block-diagonal, and so on) leads to this or that way of constructing the mapping $T \in \mathcal{F}_M$.

1.1 Schemes of parallelization

Consider Scheme 1.1.

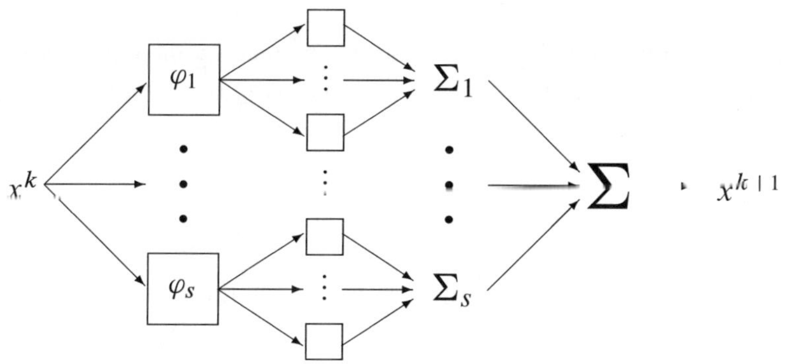

Scheme 1.1

Let the system $Ax \leq b$ be partitioned into the subsystems

$$A_j x \leq b_j, \quad j = 1, \dots, s \tag{1.1}$$

with the sets of solutions M_j, thus, $M = \{x : Ax \leq b\} = \bigcap_{j=1}^{s} M_j$. If $T_j \in \mathcal{F}_{M_j}$, then the M-Fejér mapping $T \in \mathcal{F}_M$ can be constructed, for example, by (1.3), Chapter III, with $\alpha_j = 1/s$, $T(x) = (1/s)\sum_{j=1}^{s} T_j(x)$. In Scheme 1.1, the computation of elements $x_j^k = T_j(x^k)$, $j = 1,\dots,s$, will correspond to this way; the elements are put into the summation $\sum$, obtaining the arithmetic mean $(1/s)\sum_{j=1}^{s} x_j^k = x^{k+1}$. But calculation of elements x_j^k can also be parallelized, for instance, on the account of parallelizing computations of the scalar products and norms $\|a_j\|$ used in the description of inequalities of the original systems. Such a possibility is considered in Scheme 1.1.

Consider the following example of realization of the scheme for solving the following problem of linear programming:

$$L:\ \max\{(c,x) : Ax \leq b\}.$$

The dual problem is

$$L^*:\ \min\{(b,u) : A^\top u = c,\ u \geq 0\}.$$

The pair of these mutually dual problems is equivalently reduced (Eremin, 1999) to the system

$$S: \begin{cases} Ax \leq b, \\ A^\top u = c, \quad u \geq 0, \\ (c,x) - (b,u) \geq 0. \end{cases}$$

The latter inequality in the system S can be changed by the equation $(c,x) = (b,u)$. The matrix of the system S has the following block form:

$$\begin{bmatrix} A & O \\ O & A^\top \\ c^\top & -b^\top \end{bmatrix}.$$

We consider the correspondence of this block structure to its subsystems with polyhedrons $M = \{x : Ax \leq b\}$, $M^* = \{u \geq 0 : A^\top u = c\}$ and the half-space $H = \left\{\left[\begin{smallmatrix} x \\ u \end{smallmatrix}\right] : (c,x) \geq (b,u)\right\}$. If $T \in \mathcal{F}_M$, $T^* \in \mathcal{F}_{M^*}$ and $T(z) = P_H(z)$, $z = \left[\begin{smallmatrix} x \\ u \end{smallmatrix}\right]$, then the parallelization of the Fejér process for the system S for the simultaneous solution of the problems L and L^* can be based on Scheme 1.2.

An implementation of this scheme in the analytical form will have the following representation: if $z^k = \left[\begin{smallmatrix} x^k \\ u^k \end{smallmatrix}\right]$, then, firstly, $\bar{x}^k = T(x^k)$, $\bar{u}^k = T^*(u^k)$ are calculated in parallel, and then $z^{k+1} = T_0^+(\bar{z}^k)$, $\bar{z}^k = \left[\begin{smallmatrix} \bar{x}^k \\ \bar{u}^k \end{smallmatrix}\right]$ is computed. Here, T_0^+ means the positive cut-off function along the second component of

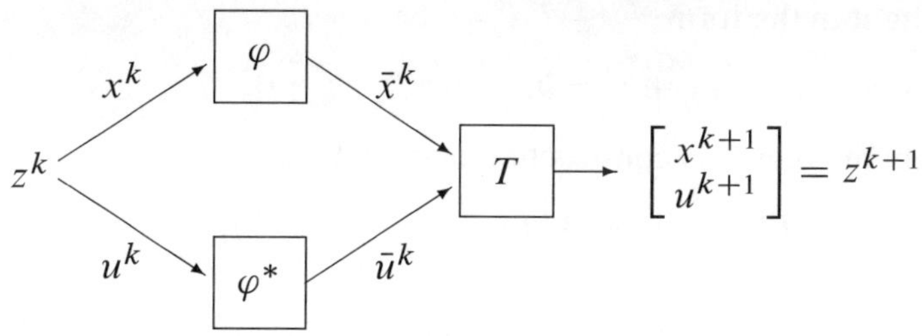

Scheme 1.2

the vector $T_0(\bar{z}^k)$. The concluding Fejér operator generating the sequence z^k can be written in the following short form:

$$\Psi(z) = T_0^+(T(x), T^*(u)).$$

The system S can be reduced to the form of a system of equations with partially nonnegative variables:

$$Ax + v = b, \quad A^\top u = c, \quad (c, x) - (b, u) = 0, \quad [v, u] \geq 0.$$

In connection to this and keeping in mind the general interest of this chapter, we separately consider the parallelization of the Fejér process for a system of the form

$$Ax = b, \quad x \geq 0. \tag{1.2}$$

We do not consider this notation in the sense of A and b used in the previously considered context (inequality systems, the linear programming problem, and so on). Let $x \in \mathbb{R}^n$, and let the rows $\{a_j\}_1^m$ of the matrix A be linear independent. Then the matrix $AA^\top$ of size $m \times m$ will be nonsingular, therefore, $(AA^\top)^{-1}$ exists. If $H = \{x : Ax = b\}$, then for the construction of the M-Fejér mapping for $M = H \cap \mathbb{R}^n_+$, one can use the known projection operator onto H:

$$P_H(x) = x - A^\top(AA^\top)^{-1}(Ax - b). \tag{1.3}$$

This operator is H-Fejér, and $T(x) = P_H^+(x)$ is M-Fejér. Since $T(x)$ is continuous, we have

$$\{T^k(x^0)\} \to x' \in M,$$

i.e., the process generated by the mapping $T(x)$ converges to a solution of system (1.2). If system (1.2) has the form

$$Ax \leq b, \quad x \geq 0, \tag{1.4}$$

then rewriting it in the form

$$Ax + v = b, \quad x \geq 0, \ v \geq 0, \tag{1.5}$$

one can apply the previous construction and obtain

$$T(z) = [z - \bar{A}^\top (\bar{A}\bar{A}^\top)^{-1}(\bar{A}z - b)]^+, \tag{1.6}$$

where $z = \left[\begin{smallmatrix} x \\ v \end{smallmatrix}\right]$, $\bar{A} = \left[\begin{array}{c|ccc} A & 1 & & \\ & & \ddots & \\ & & & 1 \end{array}\right]$, here, as everywhere in the book, the symbol $^\top$ over $\bar{A}$ means transposition. If $\{T^k(z^0)\} \to [\ x'v'\ ]$, then x' is a solution of system (1.4).

Note that in (1.4) it is not necessary to require linear independence of rows of the matrix A, and in system (1.5) this is provided automatically; this fact provides existence of the inverse matrix $(\bar{A}\bar{A}^T)^{-1}$ in formula (1.6).

Parallelization of computations of the projection operator $T(x)$ for system (1.2) with projection onto a linear manifold can be implemented on the basis of a partition of type (1.1). Namely, take an arbitrary partition

$$A_j x = b_j, \quad j = 1, \dots, k \tag{1.7}$$

of system (1.2) into subsystems, and consider

$$T_j(x) = x - A_j^\top (A_j A_j^\top)^{-1}(A_j x - b_j).$$

These are projection operators onto the manifolds $H_j = \{x : A_j x = b_j\}$, $j = 1, \dots, k$. From $\{T_j(x)\}$ it is possible to construct in many ways the operator $T \in \mathcal{F}_M$, where $M = \{x \geq 0 : Ax = b\}$. For example,

$$T(x) = \Big(\sum_{j=1}^{k} \alpha_j T_j(x)\Big)^+,$$

where $\alpha_j > 0$, $\sum_{j=1}^k \alpha_j = 1$. A parallelization of the process $\{T^k(x^0)\}$ on the basis of a partition of the original system of equations into subsystems corresponds to Scheme 1.2.

1.2 Schemes of parallelization for a linear programming problem

We next consider implementation of the parallelization Scheme 1.2 which corresponds to a partition of the system of constraints into subsystems, both in the direct and dual problems:

$$L: \ \max\{(c, x) : Ax \leq b, \ x \geq 0\},$$

$$L^*: \ \min\{(b, u) : A^\top u \geq c, \ u \geq 0\}.$$

Rewrite them in the following form,

$$\max\{(c,x):\ Ax+v=b, x\geq 0,\ v\geq 0\},$$
$$\min\{(b,u):\ A^{\top}u-w=c,\ u\geq 0,\ w\geq 0\}.$$

In this situation, the reduction $\{L, L^*\} \xrightarrow{\sim} S$ leads to the system

$$\left.\begin{aligned} Ax+v=b, \quad A^{\top}u-w=c,\\ (c,x)=(b,u),\\ x\geq 0, \quad v\geq 0, \quad u\geq 0, \quad w\geq 0.\end{aligned}\right\} \tag{1.8}$$

We now carry out a partition (in an arbitrary way, or in some other way reasonable in concrete situations) of the matrix A into horizontal A_j, $j=1,\dots,k$, and vertical H_i, $i=1,\dots,l$, submatrices. Intersections of the submatrices A_j and H_i are denoted by A_{ji}. These partitions lead to partitions of the vectors c, x, w and b, u, v into subvectors c_i, x_i, w_i and b_j, u_j, v_j, correspondingly. With these notations, system (1.8) is rewritten in the following form:

$$\sum_{i=1}^{l} A_{ji}x_i+v_j=b_j, \quad j=1,2,\dots,k; \tag{1.9$_1$}$$

$$\sum_{j=1}^{k} A_{ji}^{\top}u_j-w_i=c_i, \quad i=1,2,\dots,l; \tag{1.9$_2$}$$

$$\sum_{i=1}^{l}(c_i,x_i)=\sum_{j-1}^{k}(b_j,u_j), \tag{1.9$_3$}$$

$$x_i\geq 0, \quad v_j\geq 0; \quad u_j\geq 0, \quad w_i\geq 0. \tag{1.9$_4$}$$

It is possible to put the manifold H_j' (the set of solutions of system $(1.9)_1$) into correspondence to each $j=1,\dots,k$, and to put the manifold H_i'' (the set of solutions of system $(1.9)_2$) in correspondence to each $i=1,\dots,l$. Denoting by H_0 the hyperplane with equation $(1.9)_3$, we select (constructively) the mappings

$$T_j\in\mathcal{F}_{H_j'}, \quad T_i^*\in\mathcal{F}_{H_i''}, \quad T_0\in\mathcal{F}_{H_0}. \tag{1.10}$$

The role of these mappings can be played, in particular, by projections onto the corresponding manifolds. Here, the fact is important that in the numerical realization of the operators (1.10), the parallelization of computations is obtained by the considered system of partitions of matrices and vectors into submatrices and subvectors. In other words, the elements of parallelization are inserted into the numerical realization of the operators (1.10), for example, in the parallelization of such operations as (a_j, x), and others. Composition of all the mappings (6.10) into a sinlge one can be carried out, for instance, by the following formula:

$$T(x, u, v, w) = P_{H_0}^{+}\left(\sum_{j=1}^{k} \alpha_j T_j(x, v), \sum_{i=1}^{l} \beta_i T_i^*(u, w) \right).$$

Here, $\alpha_j > 0$, $\sum_{j=1}^{k} \alpha_j = 1$, $\beta_i > 0$, $\sum_{i=1}^{l} \beta_i = 1$; the symbol "+" above $P_{H_0}^{+}$ means the positive cut-off function on the whole system of variables $(1.9)_4$. The operator T is continuous (assuming continuity of the operators (1.10)), and $T \in \mathcal{F}_M$, where M is the set of solutions of system (1.8). Therefore,

$$T^k(x) \longrightarrow z' = \begin{bmatrix} x' \\ u' \\ v' \\ w' \end{bmatrix},$$

and $x' \in \arg L$, $u' \in \arg L^*$.

2 Randomization of Fejér processes

Fejér mappings playing the role of iterative operators for systems of linear inequalities and linear programming problems are very convenient for their combined applications. Namely, we speak about the fact that if there is a finite collection $\{T_j\}_1^m \subset \mathcal{F}_M$ of M-Fejér mappings such that

$$T_j^k(x^0) \longrightarrow x' \in M,$$

then for the realization of an iterative process, application of each operator in a given iteration can be chosen according to some *mixed* strategy:

$$x^{k+1} = \begin{cases} T_1(x^k) & \text{— with probability } p_1, \\ \vdots & \\ T_m(x^k) & \text{— with probability } p_m. \end{cases} \tag{2.1}$$

Here, $p_j \geq 0$, $\sum_{j=1}^{m} p_j = 1$. The vector $p = [p_1, \ldots, p_m]$ is called a *mixed strategy* for the application of the operators $T_j(x)$ to generate an iterative scheme. The mixed strategy can be changed from step to step, i.e., on each iterative step k the strategy $p^k = [p_1^k, \dots, p_m^k]$ is used. Evolution of the strategy p^k can be defined by taking into account the role of each operator T_j in decreasing this or that residual function $d(x)$ for the set M. If M is the polyhedron of the system $l_j(x) \le 0$, $j = 1, \dots, m$, then the functions $\max_j l_j^+(x)$ or $\sum_j l_j^+(x)$ (or others) can play the role of the function $d(x)$.

The operator taking its value accordingly to (2.1) is denoted by $T^p(x)$. This operator is realized for some $T_{j_p}(x)$. If $p = p^k$, then we write $T^{p^k}(x) = T_{j_k}(x)$. We now introduce the following notation.

Definition 2.1. The operator $T \in \mathcal{F}_M$ is called *regular*, if it satisfies relation (5.1), Chapter III.

2.1 General theorems on convergence

Theorem 2.2. *Let the following conditions be satisfied:*

1) $\{T_j\}_1^m \subset \mathcal{F}_M$, *and at least one operator from this set is either regular or continuous.*

2) *The elements of the sequence p^k of mixed strategies are separated from zero by the same distance (i.e., $\exists\, \varepsilon > 0 : p_j^k \ge \varepsilon \ \forall\, j,\ \forall\, k$).*

Then the sequence x^k given by the recurrence relation

$$x^{k+1} = T^{p^k}(x^k) \quad (= T_{j_k}(x^k)), \tag{2.2}$$

converges to $x' \in M$.

Proof. Let one of the operators $\{T_j\}_1^m$ be regular, say, $T_1(x)$. By virtue of the condition $p_1^k \ge \varepsilon > 0$, there evidently exists a sequence $x^{t_k} \to x'$, such that $x^{t_k+1} = T_1(x^{t_k}) \to x''$. But if $x' \in M$, then $x^k \to x'$; and if $x' \notin M$, then according to condition 1 we have $|x^{t_k+1} - M| = |T_1(x^{t_k}) - M| \le \Theta |x^{t_k} - M|$, $\Theta \in (0, 1)$, and from this $|x'' - M| \le \Theta |x' - M| < |x' - M|$. It is evident that a vector $y \in M$ can be found such that $\|x'' - y\| < \|x' - y\|$ in contradiction to the property of the equal distance of x' and x'' from y (see Remark 5.6, Chapter I).

But if one of the operators $\{T_j\}_1^m$ is continuous, say, $T_1(x)$, then beginning (as in the previous case) with $x^{t_k} \to x'$, $x^{t_k+1} = T_1(x^{t_k}) \to x''$ and passing to the limit in the inequality $\|T_1(x^{t_k}) - y\| \le \|x^{t_k} - y\|$, $y \in M$, we obtain $\|x'' - y\| = \|T_1(x') - y\| \le \|x' - y\|$. In this situation, if $x' \notin M$, we obtain the strict inequality $\|x'' - y\| < \|x' - y\|$ which is contradictory.

The theorem is proved. □

We next present two additional theorems on the convergence of randomized processes.

Let $\{T_j\}_{j=1}^m \subset \mathcal{F}_M$, i.e., a finite set of M-Fejér mappings be given. We then construct the mapping $T^p(x)$ according to (2.1), i.e.,

$$T^p(x) = \begin{cases} T_1(x) & \text{— with probability } p_1, \\ & \vdots \\ T_m(x) & \text{— with probability } p_m. \end{cases} \tag{2.3}$$

Here, $p_j \geq 0$ and $\sum_{j=1}^m p_j = 1$.

Theorem 2.3. *If at least one of the mappings $\{T_j\}_1^m \subset \mathcal{F}_M$, say, $T_1(x)$, is closed, and if $p_1 > 0$, then*

$$\{x^{k+1} \in T^p(x^k)\}_{k=0}^\infty \to x' \in M. \tag{2.4}$$

Proof. Note, firstly, that $T^p \in \mathcal{F}_M$. Further, if for some $k = \bar{k} : x^{\bar{k}} \in M$, then process (2.4) terminates and, therefore, the statement on convergence (2.4) is valid , i.e., $x^k = x^{\bar{k}}\ \forall k > \bar{k}$, and $x^{\bar{k}} = x'$. Now let $M \cap \{x^k\} = \varnothing$, then this sequence is M-Fejér according to the definition. Since $p_1 > 0$, the subsequence $\{x^{j_k}\}$ can be selected from the sequence $\{x^k\}$, and this subsequence satisfies $x^{j_k+1} \in T_1(x^{j_k})$ and moreover both $\{x^{j_k}\}$ and $\{x^{j_k+1}\}$ converge, say, to x' and x'', correspondingly. Then, by virtue of the closedness of the mapping $T_1(x)$, we shall have $x'' \in T_1(x')$. If $x' \in M$, then $\{x^k\} \to x'$, and then the theorem is valid. But if $x' \notin M$, then for $y \in M$ the inequality $\|x'' - y\| < \|x' - y\|$ holds which according to the property of equal distances of the limit points from y gives a contradiction. The theorem is completely proved. □

Let $T_j \in \mathcal{F}_{M_j}$, $j = 1, \dots, m$, and, let $M = \bigcap_{j=1}^m M_j \neq \varnothing$. We then construct the mappings $T_0(x) = \sum_{j=1}^m \alpha_j T_j(x), \alpha_j > 0, \sum_{j=1}^m \alpha_j = 1$ and

$$T^p(x) = \begin{cases} T_0(x) & \text{— with probability } p_0, \\ T_1(x) & \text{— with probability } p_1, \\ & \vdots \\ T_m(x) & \text{— with probability } p_m. \end{cases} \tag{2.5}$$

Here, $p_j \geq 0$, $j = 0, 1, \dots, m$, and $\sum_{j=0}^m p_j = 1$.

Theorem 2.4. *Let the mappings $\{T_j\}_{j=1}^m$ be closed and $p_0 > 0$. Then the process* (2.4) *applied with the mapping* (2.5) *converges to some point x' from M.*

Proof. The statement is proved analogically to the one of Theorem 2.3 and will be given here for completeness. To avoid bulky considerations, we shall suppose that all the mappings T_s are single-valued. This together with the condition of their closedness gives their continuity (by taking into account their Fejér property). Note, firstly, that the sequence x^k generated by mapping (2.5) is M-Fejér if $\{x^k\} \cap M = \varnothing$. We shall suppose that possible duplications are excluded from this sequence. If for some $k = \bar{k}$ we have $x^{\bar{k}} \in M = \bigcap_{s=1}^{m} M_s$, then process (2.4) evidently terminates, and the theorem is valid. So, we shall assume that $\{x^k\} \cap M = \varnothing$. The condition $p_0 > 0$ allows to choose a subsequence $\{x^{j_k}\} \subset \{x^k\}$ such that $x^{j_k} \to x'$, $x^{j_k+1} = T_0(x^{j_k}) = \sum_{s=1}^{m} \alpha_s T_s(x^{j_k}) \to x''$, and then $T_s(x^{j_k}) \to x''_s$, therefore,

$$x'' = \sum_{s=1}^{m} \alpha_s T_s(x'). \tag{2.6}$$

All these relations were written by using the M-Fejér properties of the sequence x^k, as it was already noted above. Moreover, if $x' \in M$, then $x^k \to x'$ (see Corollary 5.5, Chapter I); therefore, the statement to be proved is valid. Now let $x' \notin M$. It is evident then that the relation $s = \bar{s} : x' \notin M_{\bar{s}}$ holds at least once, and therefore the strict inequality $\|T_{\bar{s}}(x') - y\| < \|x' - y\|$, $y \in M$ holds. Taking into account this inequality and relation (2.6), we have:

$$\begin{aligned} \|x'' - y\| &= \Big\| \sum_{s=1}^{m} \alpha_s T_s(x') - y \Big\| = \Big\| \sum_{s=1}^{m} \alpha_s (T_s(x') - y) \Big\| \\ &\le \sum_{s=1}^{m} \alpha_s \|T_s(x') - y\| < \sum_{s=1}^{m} \alpha_s \|x' - y\| = \|x' - y\|. \end{aligned}$$

But since all the limit points of the M-Fejér sequence x^k have the same distance from y, we obtain a contradiction. The theorem is completely proved. □

2.2 Some notes on the realization of the processes

As it was already noted, modifications of a mixed strategy can be implemented in dependence of the "input" of each $T_j(x)$ into decreasing the residual $d(x)$. We now consider several ways for the calculation of p^k.

Let $T_j(x^k) = x_j^k$, $\delta_j^k = [d(x^k) - d(x_j^k)]^+$, where $d(x)$ is the residual function of the original system. Here, the cut-off function is taken since that in the

Fejér processes, the monotone decay of the residual function is not guaranteed. It is possible to take

$$p_j^{k+1} = \frac{\delta_j^k}{\sum_{j=1}^m \delta_j^k}. \tag{2.7}$$

The construction of p^{k+1} with the coordinates p_j^{k+1} can be carried out on the basis of the overview of the situation on the "pre-history" interval $\{x^k, \dots, x^t\}$ considered in (2.7),

$$\delta_j^k = \sum_{s=k}^{t} [\, d(x^s) - d(x_j^s)\,]^+.$$

By using in the implementations of the Fejér processes the mentioned method of control of a mixed strategy, numerical experiments were carried out. As a rule, beginning from some k', all the mappings $T_j(x)$ were rejected except one, i.e.,

$$T^{p^k}(x^k) = T_{j'}(x^k), \quad k \geq k'.$$

In such a situation it is reasonable to suppose that the operator $T_{j'}(x)$ was the most effective. We now make another observation. Additional probational computations of the intermediate estimates of effectiveness for each $T_j(x)$ require their own computational expenditures which must be taken into account. However, considering low expenditure of computations of the values $T_j(x)$ and $d(x)$ (in the class of the iterative mappings under consideration), for small m (say, $2 \leq m \leq 4$) the suggested method of selection can be considered as completely reasonable.

The construction of a trajectory $\{p^k\}$ of mixed strategies (in the implementation of the Fejér process) can be carried out also on the basis of the Brown–Robinson approach (Robinson, 1961), suggested by the authors for solving matrix games in mixed strategies. In application to our situation, the algorithm of recalculation of the strategy consists in the following. If p^k in step k is already implemented, and $T^{p^k}(x^k) = T_{j_{k+1}}(x^k)$, then p^{k+1} is calculated by the formula

$$p^{k+1} = \frac{k}{k+1}\, p^k + \frac{1}{k+1}\, e_{j_{k+1}} \quad \Big(= \frac{1}{k+1} \sum_{s=1}^{k+1} e_{j_s} \Big).$$

Here, $e_{j_{k+1}} = [\,0, \dots, 1, \dots, 0\,]$ is the unit vector of the space $\mathbb{R}^n$ with the unit at the place j_{k+1}, which corresponds to the choice of a clear strategy according to the relation $T^{p^k}(x^k) = T_{j_{k+1}}(x^k)$. The idea of the hybrid (combined) way of using several iterative mappings is itself rather positive.

3 Fejér processes and inconsistent systems for linear inequalities

Some types of M-Fejér mappings generate convergent sequences independently on non-emptyness or emptyness of the set $M = \{x : Ax \leq b\}$, i.e., independently of consistency or inconsistency of the system of inequalities under consideration. If the mapping $T \in \mathcal{F}_M$ is continuous and $M = \varnothing$, and if in addition $T^k(x^0) \to x'$ holds, then evidently, $x' = T(x')$. Therefore, in this case, the set $\widetilde{M}$ of fixed points for $T(x)$ is nonempty. It is natural to assume that the set $\widetilde{M}$ plays the role of some approximative set for the inconsistent system of inequalities.

The Fejér mappings with the outlined property can in a natural way be applied to problems of correction for inconsistent systems of linear inequalities and to improper problems (not having any solutions) of linear programming.

3.1 Preliminary notes and information

Recall that the mapping $T : \mathbb{R}^n \to \mathbb{R}^m$ is called *weakly M-Fejér*, if

$$\|T(x) - y\| \leq \|x - y\| \quad \forall\, y \in M, \quad \forall\, x \in \mathbb{R}^n, \quad \mathrm{Fix}(T) = M.$$

As it was already mentioned in Chapter I,

1) if $T(x)$ is a weak M-Fejér mapping, then for any $\lambda \in (0, 1)$

$$T_\lambda(x) = (1 - \lambda)T(x) + \lambda x, \quad T_\lambda \in \mathcal{F}_M;$$

2) if T is a nonexpansive mapping and $M = \{x : T(x) = x\} \neq \varnothing$, then $T_\lambda(x) = (1 - \lambda)T(x) + \lambda x$, $T_\lambda \in \mathcal{F}_M$;

3) the projection operator onto the convex closed set $M \subset \mathbb{R}^n$ is nonexpansive.

We shall use these properties whenever necessary (without further notice).

Lemma 3.1. *The mapping*

$$T^\lambda(x) = x - \lambda \frac{l^+(x)}{\|a\|^2} a, \quad \lambda \in (0, 2)$$

is nonexpansive. Here, $l(x) = (a, x) - b$, $a \neq 0$.

Proof. If $l(x) \leq 0$ and $l(y) \leq 0$, then $T^\lambda(x) = x$, $T^\lambda(y) = y$; therefore, the needed inequality holds. If $l(x) > 0$, $l(y) \leq 0$, then $\|T^\lambda(x) - T^\lambda(y)\| = \|T^\lambda(x) - y\| < \|x - y\|$ by virtue of $T^\lambda \in \mathcal{F}$ for $L = \{x : l(x) \leq 0\}$.

The case $l(x) \leq 0$, $l(y) > 0$ is analogous. Let $l(x) > 0$, $l(y) > 0$. Then the needed inequality follows from the following, directly verified identity:

$$\|T^\lambda(x) - T^\lambda(y)\|^2 = \|x - y\|^2 - \frac{\lambda(2-\lambda)}{\|a\|^2}[l(x) - l(y)]^2.$$

The lemma is proved. □

For the system of linear inequalities $l_j(x) = (a_j, x) - b_j \leq 0$, $j = 1, \dots, m$, with the set of solutions denoted by M (this can be empty), we use the following notations:

$$T^{(1)}(x) = x - \sum_{j=1}^{m} \lambda_j \alpha_j \frac{l_j^+(x)}{\|a_j\|^2} \cdot a_j, \tag{3.1}$$

$$T^{(2)}(x) = T_1 T_2 \dots T_m(x), \tag{3.2}$$

$$T^{(3)}(x) = x - \lambda / \delta \sum_{j=1}^{m} l_j^+(x) a_j. \tag{3.3}$$

Here, $\alpha_j > 0$, $\sum_{j=1}^m \alpha_j = 1$, $\lambda_j \in (0,2)$, $\lambda \in (0,2)$, $\delta = \sum_{j=1}^m \|a_j\|^2$, T_j, according to (1.2), Chapter III.

Lemma 3.1 implies the following corollary.

Corollary 3.2. *The mappings* (3.1)–(3.3) *are nonexpansive.*

Proof. Mappings (3.1) and (3.3) are convex combinations of nonexpansive mappings of the type $T^\lambda(x)$ from Lemma 3.1 and, therefore, evidently are nonexpansive. As for mapping (3.2), we have

$$\begin{aligned} \|T^{(2)}(x) - T^{(2)}(y)\| &\leq \|T_2 \dots T_m(x) - T_2 \dots T_m(y)\| \\ &\leq \cdots \\ &\leq \|T_m(x) - T_m(y)\| \leq \|x - y\|, \end{aligned}$$

that was claimed. □

Lemma 3.3. *Mapping* (3.3) *is Fejér with respect to the set*

$$\widetilde{M} = \arg\min_x \sum_{j=1}^{m} l_j^{+^2}(x) \ (\neq \varnothing).$$

Proof. Rewrite (3.3) in the form

$$T^{(3)}(x) = x - \frac{\lambda}{2\delta}\nabla\sum_{j=1}^{m} l_j^{+^2}(x). \tag{3.4}$$

It is seen from this that $\widetilde{M} = \{x : T^{(3)}(x) = x\}$, i.e., $\widetilde{M}$ is the set of fixed points for $T^{(3)}(x)$. For the sequel, take $\alpha \in (0,1)$ such that $\alpha^{-1}\lambda = \lambda_0 \in (0,2)$. Then $T^{(3)}(x)$ can be rewritten in the form

$$T^{(3)}(x) = \alpha T_0^{(3)}(x) + (1-\alpha)x,$$

where $T_0^{(3)}(x) = x - \frac{\lambda_0}{\delta}\sum_{j=1}^{m} l_j^+(x)a_j$. By Lemma 3.1 and Corollary 3.2 the mapping $T_0^{(3)}(x)$ is nonexpansive, and, therefore, we obtain $T^{(3)} \in \mathcal{F}_{\widetilde{M}}$ that was claimed. □

Lemma 3.4. *Mapping* (3.1)*, i.e.,*

$$T^{(1)}(x) = x - \sum_{j=1}^{m}\alpha_j\lambda_j\frac{l_j^+(x)}{\|a_j\|^2}a_j, \tag{3.5}$$

where $\lambda_j \in (0,2)$, $\alpha_j > 0$, $\sum_{j=1}^{m}\alpha_j = 1$, *is Fejér with respect to the set*

$$\widetilde{M} = \arg\min_x \sum_{j-1}^{m}\alpha_j\lambda_j\frac{l_j^{+^2}(x)}{\|a_j\|^2} \quad (\neq \varnothing).$$

Proof. This proof is analogous to the previous one. □

3.2 Fejér processes for problems of square approximation of inconsistent systems of linear inequalities

Theorem 3.5. *The process* $\{T^k(x^0)\}_{k=0}^{\infty}$ *for* $T(x)$ *of the form* (3.3) *or* (3.5) *converges to* $x' \in \arg\min_x d(x)$ *with* $d(x) = \|(Ax-b)^+\|^2$ *and* $d(x) = \sum_{j=1}^{m}\alpha_j\lambda_j\frac{l_j^{+2}(x)}{\|a_j\|^2}$, *correspondingly.*

Proof. This follows from the continuity of mappings (3.3) and (3.5) and the $\widetilde{M}$-Fejér property. □

If the system of linear inequalities

$$l_j(x) = (a_j, x) - b_j \le 0, \quad j = 1, 2, \ldots, m$$

with the set of solutions M, possibly empty, is given on an arbitrary Hilbert space $\mathcal{H}$, then it is possible to construct weakly convergent processes for approximatively solving problems on the basis of the correcting multipliers method (Section 4, Chapter I). Namely, the following statement is valid.

Theorem 3.6. *Let the operator T acting in the Hilbert space $\mathcal{H}$ be defined by formula* (3.1) *(or* (3.3)*). Let γ_k be an admissible sequence in the sense of Definition 4.4, Chapter I. Then the sequence x^k generated by the process*

$$x^{k+1} = \gamma_k T(x^k) + (1 - \gamma_k) v^0,$$

converges strongly to v^0, this is, the normal solution of the problem

$$\min_x \sum_{j=1}^{m} l_j^{+^2}(x)$$

or

$$\min_x \sum_{j=1}^{m} \alpha_j \lambda_j \frac{l_j^{+^2}(x)}{\|a_j\|^2},$$

respectively.

Proof. This follows directly from Theorem 4.5, Chapter I, Corollary 3.2 and Lemmas 3.3 and 3.4 (this chapter). □

Now we consider generalizations of the mappings $T(x)$ from Theorem 3.5 that are related to a mixed system of linear inequalities and equations:

$$\begin{aligned} l_j(x) &= 0, \qquad j \in J_=, \\ l_j(x) &\le 0, \qquad j \in J_\le, \end{aligned} \tag{3.6}$$

where $l_j(x) = (a_j, x) - b_j$, $j \in J_= \cup J_\le$. The function of the square (more exactly, piecewise square) residual of this system will have the form

$$d(x) = \sum_{j \in J_=} l_j^2(x) + \sum_{j \in J_\le} l_j^{+^2}(x). \tag{3.7}$$

The problem $\min d(x)$ is a problem of a piecewise square approximation of system (3.6). This problem is important itself, but such problems also result from linear programming problems, square-regularized problems, square programming problems, and others. For (3.6) we now consider a mapping $T(x)$ which is an analogue of (3.3):

$$T(x) = x - \frac{\lambda}{\delta}\Big[\sum_{J_=} l_j(x)a_j + \sum_{J_\leq} l_j^+(x)a_j\Big] \tag{3.8}$$

$(= x - \frac{\lambda}{2\delta}\nabla d(x))$; here, $\delta = \sum_J \|a_j\|^2$, $J = J_= \cup J_\leq$.

Let $\widetilde{M} = \arg\min d(x)$. In complete correspondence with the proof of Lemma 3.4, we can prove the following lemma.

Lemma 3.7. *Mapping* (3.8) *is continuous* $\widetilde{M}$*-Fejér, and, hence,*

$$T^k(x^0) \to x' \in \widetilde{M}.$$

□

3.3 Transition of results to the case of a system with additional constraints

For the system (3.6) with the additional constraint $x \in S$, where S is some convex closed set from $\mathbb{R}^n$, we now consider the problem

$$\min_{x \in S} d(x), \tag{3.9}$$

where $d(x)$ is a convex differentiable function. Here, for example, function (3.7) can play the role of $d(x)$.

Lemma 3.8. *The vector* $\bar{x} \in S$ *is optimal for the solvable problem* (3.9) *if and only if*

$$P_S(\bar{x} - \gamma\nabla d(\bar{x})) = \bar{x}, \quad \gamma > 0. \tag{3.10}$$

Proof. Exclude, as trivial, the case $\nabla d(\bar{x}) = 0$ (this corresponds to the case when $\bar{x}$ is already the point of absolute minimum of the function $d(x)$). Let $\bar{x} \in \widetilde{S} = \arg(3.9)$. The hyperplane $H = \{x : (\nabla d(\bar{x}), x - \bar{x}) = 0\}$ will be the support for S at the point $\bar{x}$, and then, $(\nabla d(\bar{x}), x - \bar{x}) \geq 0 \ \forall\, x \in S$. The point $\bar{z} = \bar{x} - \gamma\nabla d(\bar{x})$ has just the same point, namely, $\bar{x}$ as its projections onto S and onto H, and this corresponds to (3.10). Now conversely, let (3.10) be satisfied. Then we have the situation of a supporting hyperplane H such that the inequality $(\nabla d(\bar{x}), x - \bar{x}) \geq 0$ is satisfied for all $x \in S$. Since the function $d(x)$ is convex, the inequality $(\nabla d(\bar{x}), x - \bar{x}) \leq d(x) - d(\bar{x})$ holds for all x. This, together with the inequality written above, gives $d(x) \geq d(\bar{x})$ for all $x \in S$, i.e., $\bar{x} \in \widetilde{S}$. □

Now consider the operator $T(x)$ from (3.8) and define

$$T_\alpha(x) = (1-\alpha)\, P_S\,[\,T(x)\,] + \alpha x, \quad \alpha \in (0,1).$$

Theorem 3.9. *The mapping $T_\alpha(x)$ is continuous and $\widetilde{S}$-Fejér, where $\widetilde{S} =$* arg (3.9). *From that it follows*

$$T_\alpha^k(x^0) \to x' \in \widetilde{S}.$$

Proof. The operators T and P_S are nonexpansive and, therefore, the operator T_α is also nonexpansive, and in addition, arg (3.9) $= \mathrm{Fix}(T_\alpha) = \widetilde{S}$. By virtue of the continuity of T_α and $T_\alpha \in \mathcal{F}_{\widetilde{S}}$, we obtain the claimed result. □

4 Fejér processes for finding quasi-solutions of improper problems of linear programming (IP LP)

Let

$$L:\ \max\{(c,x)\ :\ Ax \le b,\ x \ge 0\} \tag{4.1}$$

be the linear programming problem, and let

$$L^*:\ \min\{(b,u)\ :\ A^\top u \ge c,\ u \ge 0\} \tag{4.2}$$

be the problem dual to L.

In correspondence to the problems L and L^*, we consider the following system of linear inequalities

$$S: \quad \left.\begin{array}{ll} Ax \le b,\ x \ge 0; & (4.3)_1 \\ A^\top u \ge c,\ u \ge 0; & (4.3)_2 \\ (c,x) = (b,u), & (4.3)_3 \end{array}\right\} \tag{4.3}$$

which is called *symmetric*.

In linear programming theory, the following fact is well known:

$$\boxed{\arg S = \arg L \times \arg L^*.} \tag{4.4}$$

In (4.4), arg S is the set of solutions of the system S. If

$$T_1 \in \mathcal{F}_{M_1}, \quad M_1 = \arg (4.3)_1,$$

$$T_2 \in \mathcal{F}_{M_2}, \quad M_2 = \arg (4.3)_2,$$

$$T_3(x,u) = P_{H_0}(x,u), \quad H_0 = \arg (4.3)_3$$

and

$$T(x,u) = P_H(T_1^+(x), T_2^+(u)), \tag{4.5}$$

then $T \in \mathcal{F}_{\widetilde{M}}$, where $\widetilde{M} = \arg S$, i.e., $T(x,u)$ is a continuous Fejér mapping with respect to the set $\arg L \times \arg L^*$. Therefore,

$$T^k(x_0, u_0) \to [\bar{x}, \bar{u}] \in \arg S. \tag{4.6}$$

The choice of the mappings $T_1(x)$ and $T_2(u)$ can be carried out in various ways, particularly, by the *basic* ways described in Section 1, Chapter III.

Let

$$\left.\begin{aligned} T_1(x) &= x - (\lambda / \delta_1) \sum_{j=1}^{m} l_j^+(x) a_j, \\ T_2(u) &= u - (\lambda / \delta_2) \sum_{i=1}^{m} h_i^+(u) h_i, \\ T_3(x,u) &= [x,u]^\top - \frac{(c,x)-(b,u)}{|c|^2 + |b|^2}[c,-b]^\top, \end{aligned}\right\} \tag{4.7}$$

where $\{h_i\}$ are the rows of the matrix A, $h_i(u) = c_i - (h_i, u)$, and $\delta_1 = \sum_{j=1}^{m} \|a_j\|^2$, $\delta_2 = \sum_{i=1}^{n} \|h_i\|^2$, $\lambda \in (0,2)$, $[x,u]^\top = \left[\begin{smallmatrix} x \\ u \end{smallmatrix}\right]$, $[c,-b]^\top = \left[\begin{smallmatrix} c \\ -b \end{smallmatrix}\right]$.

Note that the form of the mappings T_1 and T_2 depends on the form in which the original linear programming problem is written. For example, if the problem L is given in the following canonical form,

$$\min\{(c,x) : Ax = b,\ x \geq 0\}, \tag{4.8}$$

then the dual problem will have the form

$$\max\{(b,u) : A^\top u \leq c\}, \tag{4.9}$$

and the mappings T_1, T_2, and $T(x,u)$ have the following corresponding forms,

$$\left.\begin{aligned} T_1(x) &= x - (\lambda / \delta_1) \sum_{j=1}^{m} l_j(x) a_j, \\ T_2(u) &= u - (\lambda / \delta_2) \sum_{j=1}^{m} [(h_i, u) - c_i]^+ h_i, \\ T_3(x,u) &= P_{H_0}(T_1(x), T_2^+(u)). \end{aligned}\right\} \tag{4.10}$$

The formula of $T_3(x,u)$ contains projection operators onto H_0.

From the fact that the main constraints in (4.8) are written in the form of a system of linear equations $Ax = b$ (let us denote its set of solutions by H_1) it follows that T_1 can be constructed on the basis of the projection operator onto H_1:

$$P_{H_1}(x) = x - A^{\top}(AA^{\top})(Ax - b) \tag{4.11}$$

(under the assumption that rank $A = n$). If we assume $T_1(x) = P_{H_1}(x)$, then the final iterative operator $T(x,u)$ will have the same form as is in (4.10).

For problem (4.8) the case of a mapping $T(x,u)$ of the form

$$T(x,u,v) = P_{H_0}(x,u^+,v^+), \tag{4.12}$$

is of interest; here H_0 denotes the set of solutions of the system

$$Ax = b, \quad A^{\top}u + v = c, \quad (c,x) = (b,u). \tag{4.13}$$

This case was considered in the work (Eremin and Popov, 2002) devoted to numerical experiments for solving problems of linear programming of large dimension on parallel computers.

4.1 The basic approximative-Fejér process for improper linear programming problems of the first kind

We now shall discuss some unsolvable problems of linear programming. We start with some classifications for them.

Let M and M^* be the admissible sets of the problems L and L^*. The problem L is called an *improper problem* of the first, the second, or the third kind if the respective condition is satisfied:

1) $M = \varnothing, \quad M^* \neq \varnothing$;
2) $M \neq \varnothing, \quad M^* = \varnothing$;
3) $M = \varnothing, \quad M^* = \varnothing$.

The case $M \neq \varnothing$, $M^* \neq \varnothing$ corresponds to solvability of the problem L.

If the original problem of linear programming is unsolvable, i.e., in the terminology introduced above it is *improper*, then its corresponding symmetric system S is inconsistent (the inverse is also valid). In such a case for this problem it is possible to introduce the notion of the *quasi-solution* using the definition of the *quasi-solution* of the system S. We give an illustration by considering a linear programming problem of the form

$$L : \max\{(c,x) : Ax \le b\}. \tag{4.14}$$

Its dual problem will be

$$L^* : \min\{(b,u) : A^\top u = c,\ u \geq 0\}. \tag{4.15}$$

The symmetric system S is

$$Ax \leq b, \quad A^\top u = c, \quad u \geq 0, \quad (c,x) = (b,u). \tag{4.16}$$

If L is the improper problem of the first kind, i.e.,

$$M = \{x : Ax \leq b\} = \varnothing, \quad M^* = \{u : A^\top u = c,\ u \geq 0\} \neq \varnothing,$$

then having introduced

$$Z = \left\{ \begin{bmatrix} x \\ u \end{bmatrix} : A^\top u = c,\ (c,x) = (b,u) \right\},$$

$$d(x,u) = \|(Ax - b)^+\|^2 + \|(-u)^+\|^2,$$

it is possible to formulate the problem

$$\min \left\{ d(x,u) : \begin{bmatrix} x \\ u \end{bmatrix} \in Z \right\} \tag{4.17}$$

as an approximative problem for the inconsistent system (4.16). If $\left[\begin{smallmatrix} \bar{x} \\ \bar{u} \end{smallmatrix}\right] \in \arg(4.17)$, then $\bar{x}$ is just called the *quasi-solution* of system (4.16), and, in addition, the quasi-solution of problem (4.14). If the latter is solvable, then $\bar{x}$ is its usual solution. For (4.17) it is possible to write the Fejér mapping with respect to the set $\arg(4.17)$ in the same way as in the previous section, but in application to system (4.16) it gives:

$$\begin{aligned} T(x,u) &= \begin{bmatrix} x \\ u \end{bmatrix} - (\lambda / 2\delta) \nabla_{x,u} d(x,u), \\ \Psi(x,u) &= P_H T(x,u), \\ \Psi_\alpha(x,u) &= (1-\alpha)\,\Psi(x,u) + \alpha \begin{bmatrix} x \\ u \end{bmatrix}, \end{aligned} \tag{4.18}$$

where $\lambda \in (0,2)$, $\delta = \sum_{j=1}^m \|a_j\|^2 + m$, $\alpha \in (0,1)$.

According to Lemma 3.3, $\Psi_\alpha(x,u)$ is a $\widetilde{M}$-Fejér mapping with respect to $\widetilde{M} = \arg(4.17)$. Therefore, the process

$$\{\Psi_\alpha^k(x_0,u_0)\}_k$$

converges to some vector $\left[\begin{smallmatrix} \bar{x} \\ \bar{u} \end{smallmatrix}\right]$, and then, by definition:

– $\bar{x}$ is the quasi-solution of problem (4.14),

– $\bar{u}$ is the quasi-solution of problem (4.15).

The operator $P_H(\cdot)$ presented in the definition of $\Psi(x,u)$ and then inserted into formula (4.18), can be written according to relation (4.11), by replacing the matrix A with the matrix

$$\bar{A} = \begin{bmatrix} 0 & A^\top \\ c^\top & -b^\top \end{bmatrix},$$

and by replacing the vector b (in (4.11)) with the vector $\begin{bmatrix} c \\ 0 \end{bmatrix}$.

4.2 Approximative-Fejér process for improper problems of linear programming of the second kind

We consider this question in the context of problems (4.14)–(4.16). The situation of improperty of the second kind corresponds to the case $M \neq \varnothing$, $M^* = \varnothing$, where the sense of the symbols M and M^* is the same that in Section 3. Divide system (4.16) into two parts:

$$A^\top u = c, \quad u \geq 0; \tag{4.19}$$

$$Ax + v = b, \quad v \geq 0, \quad (c,x) = (b,u); \tag{4.20}$$

where the system $Ax \leq b$ above is changed to $Ax + v = b$, and $v \geq 0$ is a usual rewriting of the system of inequalities, if needed.

Consider $d(x,u,v) = \|A^\top u - c\|^2 + \|(-u)^+\|^2 + \|(-v)^+\|^2$, and H_0 is the set of solutions of system (4.20). By virtue of the assumption about nonemptyness of M, we have $H_0 \neq \varnothing$. In accordance with the basic construction of the Fejér mapping $\Psi_\alpha(\cdot)$ realized above for the nonlinear problems of linear programming of the first kind, in the considered case of the nonlinear problems of the second kind, this mapping will have the form

$$\Psi_\alpha(x,u,v) = P_{H_0}\left\{\begin{bmatrix} x \\ u \\ v \end{bmatrix} - (\lambda/2\delta)\nabla_{x,u,v} d(x,u,v)\right\}. \tag{4.21}$$

Here, $\lambda \in (0,2)$, $\delta = \sum_{i=1}^{n} \|h_i\|^2 + 2m$.

The process generated by mapping (4.21) will converge to some vector $[\bar{x},\bar{u},\bar{v}]$, where $\bar{x}$ is a quasi-solution of problem (4.14), and $\bar{u}$ is a quasi-solution of problem (4.15).

4.3 Fejér process for improper problems of linear programming of the third kind

Above we considered the inconsistency of the symmetric system S that was put in correspondence to the original problem of linear programming, which is supposed to be a nonlinear problem of linear programming of either the first or second kind. The sense of approximation in the case of the system S is in constructing its consistent subsystem from the equations and in introducing the square residual function $d(x)$ for the other constraints on this system. It is clear that this construction is nonunique. This follows from the fact that in any concrete case, i.e., in the case of some concrete form of the problem (for instance, in the form of a transportation problem with a block format, etc.), the mentioned operation of selecting some subsystem can be joined with some simple and effective numerical realization of the iterative step generated by this operator. It is necessary to add to this that constructing the final Fejér iterative operator will depend on the concrete form of the original problem of linear programming. All these considerations have to be taken into account each time when it is necessary to construct a Fejér mapping.

For the consideration of an improper problem of linear programming of the third kind, let us consider its dual problem of linear programming and the system S in the form (4.1)–(4.3).

Note that the final iterative mapping $\Psi_\alpha(\cdot)$ is formed by the fragments $T(x)$, $\Psi(x) = P_M T(x)$, and they, in the turn, are formed by M, $d(\cdot)$, and δ. So, when constructing the analogues of the mappings $\Psi_\alpha(\cdot)$ for the application to system (4.3) in the versions considered below, we shall present only the form of the set M, the residual function $d(\cdot)$, and the value of δ.

Version 1.

$$M_1 = \left\{ \begin{bmatrix} x \\ u \end{bmatrix} : (c, x) = (b, u) \right\},$$

$$d_1(x,u) = \|(Ax - b)^+\|^2 + \|(c - A^\top u)^+\|^2 + \|(-x)^+\|^2 + \|(-u)^+\|^2,$$

$$\delta_1 = \sum_{j=1}^{m} \|a_j\|^2 + \sum_{i=1}^{n} \|h_i\|^2 + m + n.$$

Version 2.

$$M_2 = \left\{ \begin{bmatrix} x \\ u \\ v \end{bmatrix} : A^\top u - v = c,\ (c,x) = (b,u) \right\},$$

$$d_2(x,u,v) = \|(Ax-b)^+\|^2 + \|(-x)^+\|^2 + \|(-u)^+\|^2 + \|(-v)^+\|^2,$$

$$\delta_2 = \sum_{j=1}^{m} \|a_j\|^2 + m + 2n.$$

Version 3.

$$M_3 = \left\{ \begin{bmatrix} x \\ u \\ v \end{bmatrix} : Ax - w = b,\ (c,x) = (b,u) \right\},$$

$$d_3(x,u,w) = \|(c - A^\top u)^+\|^2 + \|(-x)^+\|^2 + \|(-u)^+\|^2 + \|(-w)^+\|^2,$$

$$\delta_3 = \sum_{i=1}^{n} \|h_i\|^2 + 2m + n.$$

Version 4.

$$M_4 = \mathbb{R}^n,$$

$$d_4(x,u) = |(Ax-b)^+|^2 + |(c - A^\top u)^+|^2 + |(-x)^+|^2 + |(-u)^+|^2,$$

$$\delta_4 = \sum_{j=1}^{m} \|a_j\|^2 + \sum_{i=1}^{n} \|h_i\|^2 + m + n.$$

The latter version is suitable both for solvable and unsolvable problems of any type of insolvability (of the 1st, 2nd, and 3rd kind).

5 Normalized solutions of convex inequalities

Let M be a convex closed subset of $\mathcal{H}$ and $v \notin M$.

The element $\bar{v} = \arg\min_{x \in M} \|v - x\|$ that is the metric projection v onto M is called the v-*normal* element from M. If M is given by the system of convex inequalities, say,

$$f_j(x) \le 0, \quad j = 1, \dots, m, \tag{5.1}$$

then the element $\bar{v}$, according to Definition 2.8, Chapter I, is called the v-*normal* solution of system (5.1). If $v = 0$, then $\bar{v}$ is the solution of system (5.1) having minimal norm, and it is called *normal* solution.

5.1 Auxiliary results

Lemma 5.1. *An element $\bar{v}$ from M is the metric projection of an element $v \notin M$ if and only if for any $x \in M$ the following inequality holds:*

$$(x - \bar{v}, v - \bar{v}) \le 0, \tag{5.2}$$

see, for instance, (Berdyshev and Petrak, 1999). □

We shall need an equivalent characterization of an M-Fejér mapping $T \in \mathcal{F}_M$.

Let π be the hyperplane that is the locus of the points lying at the same distance form x and $T(x)$; the equation of this hyperplane has the form

$$(y - \bar{x}, x - T(x)) = 0. \tag{5.3}$$

Here, $x \notin M$, $\bar{x} = \frac{x+T(x)}{2}$, and y is a variable. By π_-^0 we denote the open half-space corresponding to the inequality

$$(y - \bar{x}, x - T(x)) < 0, \tag{5.4}$$

and by π_+^0 we denote the open half-space corresponding to the inequality

$$(y - \bar{x}, x - T(x)) > 0. \tag{5.5}$$

Lemma 5.2. *Let M be a convex closed subset of $\mathcal{H}$. The mapping T from $\mathcal{H}$ into $\mathcal{H}$ is M-Fejér if and only if $T(y) = y \ \forall\, y \in M$ and for any $x \notin M$ the following inequality holds:*

$$(y - \bar{x}, x - T(x)) < 0, \quad \forall\, y \in M. \tag{5.6}$$

Here, $\bar{x} = \frac{x+T(x)}{2}$.

Proof. Necessity. Let $T \in \mathcal{F}_M$, then it is necessary to show $M \subset \pi_-^0$, i.e., that inequality (5.6) holds for all $y \in M$. Suppose that it does not hold, i.e., for $x \notin M$ the point y belongs either to the hyperplane π or the open half-space π_+^0. If $y \in \pi$, then $\|T(x) - y\| = \|x - y\|$ (since x and $T(x)$ are at the same distances from all points of the set π), but this contradicts the Fejér property of $T(x)$ with respect to the set M, i.e., it contradicts to the property $\|T(x) - y\| < \|x - y\|$. But if $y \in \pi_+^0$, then $T(x)$ and y will lie on different sides of the hyperplane π. Denote by z the point of intersection of the segment $[T(x), y]$ and π. Then we shall have $\|T(x) - y\| = \|T(x) - z\| + \|z - y\| = \|x - z\| + \|z - y\| \ge \|x - y\|$, but this also contradicts to the inequality $\|T(x) - y\| < \|x - y\|$.

Sufficiency. Let (5.6) be satisfied together with the property $T(y) = y$ $\forall\, y \in M$. It is necessary to ensure that the inequality $\|T(x) - y\| < \|x - y\|$ for $y \in M$ holds and that $x \notin M$. The transfer from x to $T(x)$ means *relaxation* of the point x with respect to the projection of x onto π with the relaxation coefficient $\lambda = 2$, i.e., $T(x) = x - 2\,(x - P_\pi)$, that gives the equality $\|T(x) - y\| \leq \|x - y\|$. But since $y \in M \subset \pi^0_-$, i.e., y does not lie on the hyperplane π, the strict inequality $\|T(x) - y\| < \|x - y\|$ actually holds which was to be shown.

The lemma is completely proved. □

We consider the following mapping for $T \in \mathcal{F}_M$ and $v \notin M$:

$$T_\alpha(x) = (1 - \alpha)\, T(x) + \alpha v, \tag{5.7}$$

$\alpha \in (0, 1)$. Let $\bar{v}$ be the projection v onto the set M.

Lemma 5.3. *If $S = \{x : \|x - \bar{v}\| \leq r\}$ where $r = \|v - \bar{v}\|$, then*

$$T_\alpha(S) \subset S,$$

i.e., $x \in S \Longrightarrow T_\alpha(x) \in S$.

Proof. Actually, if $\|x - \bar{v}\| \leq r$, then

$$\begin{aligned}\|T_\alpha(x) - \bar{v}\| &= \|(1 - \alpha)T(x) + \alpha v - \bar{v}\| \\ &= \|(1 - \alpha)(T(x) - \bar{v}) + \alpha(v - \bar{v})\| \\ &\leq (1 - \alpha)\,\|x - \bar{v}\| + \alpha\, r \leq (1 - \alpha)\, r + \alpha r = r.\end{aligned}$$ □

Corollary 5.4. *Let the operator $T \in \mathcal{F}_M$ be completely continuous. Then by virtue of the Schauder theorem* (Edwards, 1969) *the mapping $T_\alpha(x)$ has at least one fixed point x_α,*

$$x_\alpha = (1 - \alpha)T(x_\alpha) + \alpha v. \tag{5.8}$$

□

One can weaken the conditions on the operator T which guarantee existence of the fixed points of the operator T_α. We formulate these conditions in the form of a corollary, since, here, the fact established in Lemma 5.3 $T_\alpha \subseteq S$ is also used.

Corollary 5.5. *Let the operator $T \in \mathcal{K}_M$ (Definition 1.1, Chapter I) acting in the Hilbert space $\mathcal{H}$ be weakly closed (Definition 3.4, Chapter I). Then for any $\alpha \in (0, 1)$, equation (5.8) has a solution $x_\alpha \in S$.*

Proof. Actually, write equation (5.8) in the form

$$\min_{x \in S} \|x - T_\alpha(x)\| \ (= d).$$

Let $x_n \in S$ be a minimizing sequence. Since $S = \{x : \ \|x - \bar{v}\| \leq r\}$ is bounded and $T_\alpha(S) \subseteq S$, the subsequence exists such that $x_{n_k} \rightharpoonup \bar{x} \in S$, $T_\alpha(x_{n_k}) \rightharpoonup \bar{y}$. So as T is weakly closed then T_α is also weakly closed; therefore, $T_\alpha(\bar{x}) = \bar{y}$. Then $d \leq \|\bar{x} - T_\alpha(\bar{x})\| \leq \lim_{k\to\infty} \|x_{n_k} - T_\alpha(x_{n_k})\| = d$, i.e., $\bar{x}$ is the solution of equation (5.8). □

Note that the conditions on the operator T from Corollary 5.5 *a fortiori* hold for the basic Fejér operators (3.12), Chapter I, since for them the stronger property (see relation (3.1) and Remark 3.15, Chapter I) holds.

5.2 Theorems on stability of the fixed points for quasi-contractions

In Section 4, Chapter I, for nonexpansive operators $T: \mathcal{H} \to \mathcal{H}$ (the class $\mathcal{K}$), a theorem on convergence of the sequence x_α ($\alpha = 1 - \gamma$) to the v-normal solution of the equation $x = T(x)$ was proved (Theorem 4.1, Chapter I). In this section this fact is established under some additional conditions on the mappings from the classes $\mathcal{F}_M$, $\mathcal{K}_M$ ($M = \mathrm{Fix}(T)$), and several examples are discussed.

Theorem 5.6. *Let the mapping $T : \mathcal{H} \to \mathcal{H}$, $T \in \mathcal{F}_M$, be completely continuous, i.e., from $x_k \rightharpoonup x'$, it follows $Tx_k \to Tx'$. Then*

$$x_\alpha \to \bar{v} \text{ as } \alpha \to 0, \tag{5.9}$$

where $\bar{v}$ is the projection of v onto M.

Proof. 1. Denote by $\{x_\alpha\}'_\alpha$ the set of limit points of subsequences of the form $\{x_{\alpha_k}\}_k$, where $\{\alpha_k > 0\}_k \to 0$. Select some arbitrary sequence x_{α_k} such that $x_{\alpha_k} \rightharpoonup x'$ (since in $\mathcal{H}$ from any bounded set it is possible to select some sequence that converges weakly). We prove that $x' \in M$. Rewrite relation (5.8), with the number α_k being substituted by α:

$$x_{\alpha_k} = (1 - \alpha_k)\,T(x_{\alpha_k}) + \alpha_k v, \tag{5.10}$$

$k = 1, 2, \ldots$. Since, $\alpha_k v \to 0$, $(1 - \alpha_k) \to 1$, and $\{T(x_{\alpha_k})\}$ is a strongly convergent sequence (by assumption $T(x)$ is a completely continuous mapping), we have $x_{\alpha_k} \to x'$, i.e., convergence in the norm of the space $\mathcal{H}$ holds. Passing in (5.10) to the limit as $k \to \infty$, we obtain $T(x') = x'$, i.e., $x' \in M$.

2. We prove that $x' = \bar{v}$, i.e., that the limit point of the sequence x_{α_k} coincides with the projection $\bar{v}$ of the element v onto M.

Note that $x_\alpha \notin M$ for any $\alpha \in (0, 1)$. In fact, if $x_\alpha \in M$ would hold then from (5.8) it would follow $x_\alpha = v$, i.e., $v \notin M$, a contradiction.

From Lemma 5.2 by virtue of $T \in \mathcal{F}_M$, the following inequality is valid:

$$\Big(y - \frac{x_\alpha + T(x_\alpha)}{2},\ x_\alpha - T(x_\alpha)\Big) < 0, \quad \forall y \in M. \tag{5.11}$$

Since according to (5.8) $x_\alpha - T(x_\alpha) = \alpha\,(v - T(x_\alpha))$, it is possible to write (5.11) in the form

$$\Big(y - \frac{x_\alpha + T(x_\alpha)}{2},\ v - T(x_\alpha)\Big) < 0 \quad \forall\, y \in M. \tag{5.12}$$

Having substituted $\alpha = \alpha_k$, $k = 1, 2, \ldots$, into (5.12), we obtain

$$\Big(y - \frac{x_{\alpha_k} + T(x_{\alpha_k})}{2}),\ v - T(x_{\alpha_k})\Big) < 0 \quad \forall\, y \in M. \tag{5.13}$$

As it was already noted in point 1 of this proof, there holds $x_{\alpha_k} \to x'$, $T(x_{\alpha_k}) \to x'$. Passing to the limit in (5.13) as $k \to \infty$, we obtain

$$(y - x', v - x') \le 0 \quad \forall\, y \in M,$$

but this, according to Lemma 5.1, implies that x' is the projection of v onto M, i.e., $x' = \bar{v}$. □

Thus, we have proved that any weakly convergent sequence x_{α_k} satisfying relation (5.10) converges in norm to the projection $\bar{v}$ of v onto M.

Remark 5.7. Let us consider the situation $\mathcal{H} = \mathbb{R}^n$. In that case, Corollary 5.4 and Theorem 5.6 can be formulated as follows: If $T \in \mathcal{F}_M$ and T is continuous, then for $\alpha \in (0, 1)$ the mapping $T_\alpha(x) = (1 - \alpha)T(x) + \alpha v$, $v \notin M$, has the set of fixed points $M_\alpha \neq \emptyset$, and moreover,

$$\sup_{z \in M_\alpha} \|z - \bar{v}\| \to 0 \text{ as } \alpha \to 0. \tag{5.14}$$

If in addition the operator is nonexpansive, then M_α is single-point and instead of (5.14) we have (5.9).

Remark 5.8. If we additionally assume in Theorem 5.6 that the mapping $T(x)$ is nonexpansive, i.e., $\|T(x) - T(y)\| \le \|x - y\|$ for all $x,\ y \in \mathcal{H}$, then the set M_α of fixed points of the mapping $T_\alpha(x)$ will consist of a single point x_α, and

$$\|T_\alpha(x) - x_\alpha\| \le (1 - \alpha)\,\|x - x_\alpha\| \tag{5.15}$$

for any point $x \in \mathcal{H}$.

Note that the relations (5.15) show that under the condition of nonexpansiveness of the operator $T(x)$, the fixed point x_α for $T_\alpha(x)$ can be determined with any desirable accuracy by means of the iterative process

$$x^{k+1} = T_\alpha^k(x^0) \tag{5.16}$$

that gives x_α in the limit.

Consider several examples in the case when M is given by systems of linear and convex inequalities.

1. Let a system (5.1) of convex inequalities on $\mathbb{R}^n$ with the set of solutions $M \neq \varnothing$ be given. Assume that the functions $\{f_j(x)\}_1^m$ are differentiable. Moreover let

$$T(x) = x - \lambda \sum_{j=1}^{m} \alpha_j \frac{f_j^+(x)}{\|\nabla f_j(x)\|^2} \cdot \nabla f_j(x), \tag{5.17}$$

$\lambda \in (0, 2)$, $\alpha_j > 0$, $\sum_{j=1}^m \alpha_j = 1$. In Section 7, Chapter I, it was shown that $T \in \mathcal{F}_M$. The convex differentiable functions are continuously differentiable, i.e., the gradients $\nabla f_j(x)$ are continuous. This implies the continuity of the mapping (5.17). For $T(x)$ given according to (5.17), all the conditions of Theorem 5.6 are satisfied in the case $\mathcal{H} = \mathbb{R}^n$. Therefore, all above presented results on v-normal solutions of the system (5.1) hold also for the mapping $T \in \mathcal{F}_M$.

2. Let M be given by the following finite system of linear inequalities in $\mathbb{R}^n$,

$$l_j(x) = (a_j, x) - b_j \le 0, \quad j = 1, 2, \dots, m, \tag{5.18}$$

and let the mapping $T \in \mathcal{F}_M$ be given according to the relation

$$T(x) = x - \lambda\,/\,\delta \sum_{j=1}^{m} l_j^+(x) \cdot a_j. \tag{5.19}$$

Here, $\lambda \in (0, 2)$, $\delta = \sum_{j=1}^m \|a_j\|^2$, and M is the nonempty set of solutions of system (5.18).

As it was already noted in Section 3, mapping (5.19) is continuous M-Fejér and nonexpansive, therefore Remark 5.8 can be applied to this mapping. Recall that the remark is directly related with the realization of the computational procedure for finding the v-normal solution of system (5.18). This is completely applicable to the case of system (5.18) with the additional constraint $x \in S$, where S is a convex closed subset of $\mathbb{R}^n$. In other words, to the system

$$l_j(x) \leq 0, \quad j = 1, 2, \ldots, m, \quad x \in S \tag{5.20}$$

we put in correspondence the mapping

$$T'(x) = P_S\,[T(x)]. \tag{5.21}$$

This mapping is continuous Fejér with respect to the set $M \cap S \neq \varnothing$ and nonexpansive, i.e., again all the conditions of Remark 5.8 hold, hence for the operator (5.21) relation (5.15) is fulfilled.

In the next theorem, stability of the solutions x_α of equation (5.8) w.r.t. the parameter $\alpha \to 0$ is established under weaker assumptions on the operator T than those considered in Theorem 5.6. This allows to consider some important classes of basic Fejér constructions in the case of an infinite-dimensional Hilbert space $\mathcal{H}$.

Theorem 5.9. *Let $T \in \mathcal{K}_M$ be a weakly closed operator acting in the Hilbert space $\mathcal{H}$. Then for x_α the statement of Theorem 5.6 is valid.*

Proof. According to Corollary 5.5, for any $\alpha \in (0, 1)$ there exists a solution $x_\alpha \in S$ of equation (5.8). Since S is a bounded set, one can select a subsequence $x_{\alpha_k} \rightharpoonup \bar{x} \in S$ that converges weakly. Moreover, we have

$$x_{\alpha_k} - T(x_{\alpha_k}) = \alpha_k\,(v - T(x_{\alpha_k})) \to 0, \quad \alpha_k \to 0,$$

i.e., $T(x_{\alpha_k}) \rightharpoonup \bar{x}$. Taking into account the weak closedness, we obtain $\bar{x} = T(\bar{x})$. Let $\bar{v}$ be the projection of the element v onto the set M, i.e., $\bar{v}$ is the fixed point of the operator T that is closest to v.

Consider the relations

$$\begin{aligned} \alpha_k x_{\alpha_k} + (1 - \alpha_k)\,(x_{\alpha_k} - T(x_{\alpha_k})) &= \alpha_k v, \\ \alpha_k \bar{v} + (1 - \alpha_k)\,(\bar{v} - T(\bar{v})) &= \alpha_k \bar{v}. \end{aligned}$$

Multiplying each of these equalities term-by-term with $x_{\alpha_k} - \bar{v}$ and subtracting term-by-term gives

$$\alpha_k (x_{\alpha_k} - \bar{v},\, x_{\alpha_k} - \bar{v}) + (1 - \alpha_k)\,(x_{\alpha_k} - \bar{v},\, F(x_{\alpha_k}) - F(\bar{v})) \\ = \alpha_k (v - \bar{v},\, x_{\alpha_k} - \bar{v}),$$

where $F(x) = x - T(x)$. Since $T \in \mathcal{K}_M$, i.e., $\|T(x) - z\| \leq \|x - z\|$ $\forall\, z \in M = \mathrm{Fix}(T)$, we have

$$(F(x) - F(z),\ x - z) = \|x - z\|^2 - (T(x) - T(z),\ x - z)$$
$$\geq \|x - z\|^2 - \|T(x) - T(z)\|\,\|x - z\| \geq 0.$$

Therefore, $(x_{\alpha_k} - \bar{v},\ F(x_{\alpha_k}) - F(\bar{v})) \geq 0$.

Taking this fact into account, we obtain

$$(x_{\alpha_k} - \bar{v},\ x_{\alpha_k} - \bar{v}) \leq \alpha_k(v - \bar{v},\ x_{\alpha_k} - \bar{v})$$
$$= (v - \bar{v},\ x_{\alpha_k} - \bar{x}) + (v - \bar{v},\ \bar{x} - \bar{v}).$$

By the criterion of the metric projection, the second term on the right-hand side is nonpositive, and the first term tends to zero as $\alpha_k \to 0$ by virtue of the weak convergence $x_{\alpha_k} \rightharpoonup \bar{x}$; therefore,

$$\lim_{k \to \infty} \|x_{\alpha_k} - \bar{v}\| = 0.$$

Since it follows from the proof that $\bar{v}$ is the unique limit point, the whole sequence x_α converges to $\bar{v}$. □

As it was already mentioned above, the operators T for the basic Fejér constructions, for instance of form (3.12), Chapter I, have the weak closedness property under the condition of the boundedness of the subdifferential; however, complete continuity of the operator T cannot be satisfied.

5.3 Iterative procedure for finding projection

Below we consider an iterative procedure for finding the v-normal element from $M \subset \mathbb{R}^n$ under the assumption of nonexpansiveness of the mapping $T \in \mathcal{F}_M$. Example (5.19) and the example from Theorem 6.1 (see below) can be illustrations of such a situation. The fixed point x_α of the mapping $T_\alpha(x)$ is the unique solution of the equation

$$x = (1 - \alpha)\,T(x) + \alpha v, \tag{5.22}$$

so, the measure of vicinity of x to x_α can be defined by the term

$$d_\alpha(x) = \|x - T_\alpha(x)\|.$$

Since for $T_\alpha(x)$, relation (5.15) holds, i.e.,

$$\|T_\alpha(x) - x_\alpha\| \leq (1 - \alpha)\,\|x - x_\alpha\|,$$

the process $\{T_\alpha^k(x_0)\}_k$ for arbitrary initial x_0 gives x_α as the limit. Therefore, for sufficiently large t, the following inequality is obtained,

$$d_\alpha(x_0^t) \le \varepsilon, \quad \text{where} \quad x_0^t = T_\alpha^t(x_0), \tag{5.23}$$

where ε is an arbitrary small positive number.

Take $\alpha \in (0,1)$, $\delta > 0$, and $\varepsilon > 0$ satisfying the relation

$$\varepsilon\,/\,\alpha \le \delta. \tag{5.24}$$

Lemma 5.10. *Let α, ε, and δ satisfy condition* (5.24) *and $d_\alpha(x) \le \varepsilon$. Then*

$$\|x - x_\alpha\| \le \delta,$$

where x_α is a unique solution of (5.22).

Proof. Write (5.15) in the form

$$\|T_\alpha(x) - T_\alpha(x_\alpha)\| \le (1-\alpha)\,\|x - x_\alpha\|.$$

From this we obtain the following estimate:

$$\begin{aligned}\|x - x_\alpha\| &= \|(x - T_\alpha(x)) + (T_\alpha(x) - x_\alpha)\| \\ &\le d_\alpha(x) + \|T_\alpha(x) - T_\alpha(x_\alpha)\| \\ &\le \varepsilon + (1-\alpha)\,\|x - x_\alpha\|,\end{aligned}$$

i.e.,

$$\|x - x_\alpha\| \le \varepsilon\,/\,\alpha \le \delta,$$

which was to be proved. □

The computational process following from the above reasonings can be described in the following form:

1) consider sequences $\{\alpha_k > 0\} \to 0$, $\{\delta_k > 0\} \to 0$ and $\{\varepsilon_k > 0\} \to 0$, such that for each k relation (5.24) holds, i.e.,

$$\varepsilon_k\,/\,\alpha_k \le \delta_k, \quad k = 1, 2, \ldots; \tag{5.24$_k$}$$

2) take an initial approximation x^0;
3) if x^{k-1} has been already computed, then find $\bar{t}$ such that

$$d_{\alpha_k}(x_k^{\bar{t}}) \le \varepsilon_k, \quad \text{where } x_k^{\bar{t}} = T_{\alpha_k}^{\bar{t}}(x_k)$$

(that is possible by virtue of (5.23));

4) take $x^k = x_k^{\bar{t}}$.

The following theorem is valid.

Theorem 5.11. *The sequence generated according to rules 1–4 converges to the projection $\bar{v}$ of v onto M.*

Proof. This follows from the fact that $x_{\alpha_k} \to \bar{v}$ as $\alpha_k \to 0$ (Theorem 5.6) and from the relations

$$\|x^k - x_{\alpha_k}\| \le \delta_k, \quad k = 0, 1, 2, \dots$$

(Lemma 5.10). □

Remark 5.12. The choices $\alpha_k = \delta_k = k^{-1}$, $\varepsilon_k = k^{-2}$ provide an example that satisfies $(5.24)_k$. Another example is $\alpha_k \to 0$, $\delta_k = \alpha_k$, $\varepsilon_k = \alpha_k^2$, or $\alpha_k \to 0$, $\delta_k \to 0$, $\varepsilon_k \le \alpha_k\, \delta_k$.

Remark 5.13. The computational process described above is a modified version of the method of correcting multipliers (see Subsection 4.2, Chapter I). The difference is in the fact that here it is not necessary to put any *a priori* conditions on the admissible sequence ($\gamma_k = 1 - \alpha_k$), and the convergence is achieved on the account of the appropriate choice of the number of iterations that agrees with the value of correction and the parameter α.

6 Fejér processes for inconsistent linear and convex inequality systems

Firstly, we shall only consider the cases of geometrical formulation of the problem for finding the solutions (quasi-solutions) of the system of inclusions

$$x \in M_j, \quad j = 1, \dots, m, \tag{6.1}$$

which, not obligatory, may be consistent. Moreover, we shall be rather interested in the situation when $M = \bigcap_j M_j = \varnothing$. Below, $\{M_j\}_1^m$ are nonempty, convex, and closed subsets of $\mathbb{R}^n$.

Introduce the notations: $P_j(x)$ is the projection of x onto M_j, that is, $P_j(x) = \arg\min_{z \in M_j} \|x - z\|$. It is well known that

$$\|P_j(x) - P_j(z)\| \le \|x - z\| \quad \forall\, x,\, y \in \mathbb{R}^n, \tag{6.2}$$

i.e., $P_j(x)$ is a projection operator.

Construct the operator

$$T(x) = \sum_{j=1}^{m} \alpha_j P_j(x), \tag{6.3}$$

$\alpha_j > 0$, $\sum_{j=1}^{m} \alpha_j = 1$. Now we transform this to

$$T(x) = x - 1/2 \sum_{j=1}^{m} \nabla d_\alpha(x), \tag{6.4}$$

where $d_\alpha(x) = \sum_{j=1}^{m} \alpha_j \|x - P_j(x)\|^2$. Denoting $\rho_j(x) = \|x - P_j(x)\|$, recall the formulas

$$\nabla \rho_j^2(x) = 1/2\,(x - P_j(x)), \quad \nabla \rho_j(x) = \frac{x - P_j(x)}{\rho_j(x)}. \tag{6.5}$$

Here, the second formula is considered for $x \notin M_j$.

Introduce the condition

$$\widetilde{M} = \arg\inf_x d_\alpha(x) \neq \varnothing. \tag{6.6}$$

The following condition is, for example, sufficient for (6.6): *If for some* $\gamma \in \mathbb{R}$ *the following equality holds,*

$$M^\gamma = \{x : d_\alpha(x) \leq \gamma\} \neq \varnothing,$$

then M^γ *is bounded.* The boundedness of one of the $\{M_j\}_1^m$ is a particular case of the presented condition.

Theorem 6.1. *Let condition* (6.6) *be satisfied,* $\lambda \in (0, 1)$, *and*

$$T_\lambda(x) = (1 - \lambda)T(x) + \lambda x, \tag{6.7}$$

where T *is given by* (6.4). *Then*

1) $T_\lambda(x)$ *is a nonexpansive operator (and therefore is continuous);*

2) $T_\lambda \in \mathcal{F}_{\widetilde{M}}$.

Therefore, for an arbitrary initial $x^0 \in \mathbb{R}^n$, *the sequence* x^k *generated by the relation*

$$x^{k+1} = T_\lambda(x^k) \tag{6.8}$$

converges to $\tilde{x} \in \widetilde{M}$, *where* $\widetilde{M}$ *is given by* (6.6).

Proof. Property 1 for (6.3) follows from (6.2). Consider property 2. The function $d_\alpha(x)$ is convex and differentiable, and the following relation holds for this function:

$$\widetilde{M} = \{x : \nabla d_\alpha(x) = 0\}.$$

This relation represents the fact that the point $\bar{x}$ is a minimizer for the function $d_\alpha(x)$ if and only if $\nabla d_\alpha(\bar{x}) = 0$ holds, i.e., $\widetilde{M}$ is the set of fixed points of the operator $T(x)$. Since this operator is nonexpansive, we have $T_\lambda \in \mathcal{F}_{\widetilde{M}}$. □

Remark 6.2. The operator (6.7) can be written in the modified form

$$T_\lambda = x - (\lambda / 2)\, \nabla d_\alpha(x). \tag{6.9}$$

Remark 6.3. If $\alpha_j = 1/m$, $\forall\, j = 1, \dots, m$, then the operator (6.9) can be written in the form

$$T_\lambda^0(x) = x - (\lambda / 2m)\, \nabla d(x),$$

where $d(x) = \sum_{j=1}^m \rho_j^2(x)$.

Remark 6.4. If for system (6.1) we have $\bigcap_j M_j \neq \varnothing$, then $\bigcap_j M_j = \widetilde{M}$, and in addition the sequence x^k generated inductively by relation (6.8) will converge to an element of the intersection $\bigcap_j M_j$.

Consider the case of operator (6.3) when α_j are functions of x, namely: let $\alpha_j(x)$ be nonnegative convex functions, and then $\alpha_j(x) > 0$ for $x \notin M_j$; and $\sum_{j=1}^m \alpha_j(x) = 1$ for $x \notin \bigcap_j M_j$. Therefore, if $\bigcap_j M_j = \varnothing$, then $\sum_{j=1}^m \alpha_j(x) = 1$ for all $x \in \mathbb{R}^n$. For example, one may take

$$\alpha_j(x) = \frac{\rho_j(x)}{\sum_{s=1}^m \rho_s(x)}. \tag{6.10}$$

This is the case we shall consider. Thus, let

$$T(x) = \sum_{j=1}^m \alpha_j(x) P_j(x), \tag{6.11}$$

where $\alpha_j(x)$ is given according to (6.10). We shall suppose (to avoid special remarks) that $\bigcap_j M_j = \varnothing$. Note that $\alpha_j(x)$ may also be zero: $\alpha_j(x) = 0 \Longrightarrow x \in M_j$. Take $\delta(x) = \sum_{j=1}^m \rho_j(x)$, i.e., $\delta(x)$ is the denominator in (6.10), and

$d(x) = \sum_{j=1}^m \rho_j^3(x)$. We transform relation (6.11) by taking into account formulas (6.5):

$$\begin{aligned} T(x) &= \sum_{j=1}^m \alpha_j(x) P_j(x) \\ &= x - \sum_{j=1}^m \alpha_j(x)(x - P_j(x)) \\ &= x - \delta^{-1}(x) \sum_{j=1}^m \rho_j^2(x) \frac{x - P_j(x)}{\|x - P_j(x)\|} \\ &= x - \delta^{-1}(x) \sum_{j=1}^m \rho_j^2(x) \nabla \rho_j(x) \\ &= x - \frac{\delta^{-1}}{3} \nabla \sum_{j=1}^m \rho_j^3(x) = x - \frac{\delta^{-1}(x)}{3} \nabla d(x). \end{aligned}$$

Just as in case (6.7) construct

$$T_\lambda(x) = (1-\lambda)\left[x - \frac{\delta^{-1}(x)}{3} \nabla d(x)\right] + \lambda x \tag{6.12}$$

$(= x - \frac{1-\lambda}{3} \cdot \delta^{-1}(x) \nabla d(x))$, $\lambda \in (0,1)$.

For the case of the choice of $\alpha_j(x)$ according to (6.10) and the choice of $T(x)$ according to (6.11) (or in the modified form obtained from (6.12)), a complete analogue of Theorem 6.1 is valid in the following formulation.

Theorem 6.5. *Let* $d(x) = \sum_{j=1}^m \rho_j^3(x)$, $\delta(x) = \sum_{j=1}^m \rho_j(x)$, *and* $\widetilde{M} = \arg\inf_x d(x) \neq \varnothing$, $\lambda \in (0,1)$. *Then the mapping*

$$T_\lambda(x) = x - \frac{1-\lambda}{3} \delta^{-1}(x) \nabla d(x)$$

(i.e., (6.12)*) is nonexpansive (therefore, continuous), and moreover* $T_\lambda \in \mathcal{F}_{\widetilde{M}}$.

□

From this, convergence of the process generated by the relation

$$x^{k+1} = T_\lambda(x^k)$$

to a point from $\widetilde{M}$ follows.

Remark 6.6. If in the case of the choice of $\alpha_j(x)$ according to (6.10), the set $M = \bigcap_j M_j$ is nonempty, then $M = \widetilde{M}$; if moreover $x \in M$, i.e., $\delta(x) = 0$, then $T_\lambda(x) = x$ follows by definition.

Remark 6.7. Indeed, there are many ways for the choice of $\{\alpha_j(x)\}_1^m$ with the conditions discussed above; only in the example of (6.10), we have illustrated the techniques for constructing the mappings of type (6.11) with obtaining the theorems similar to the ones formulated above.

Bibliography

[Acar R. and Vogel C.B. (1994)] Analysis of bounded variation penalty method for ill-posed problems. *Inverse Probl.*, 10, 1217–1228.

[Ageev A.L., Vasin V.V., Bessonova E.N., and Markushevich V.M. (1997)] Radio-sounding of the ionosphere on two friquences. Algorithmic analysis of the Fredholm–Stiltjes integral equation. Theoretical problems in geophysics. *Zhurnal Vych. Seismologiya*, 29, 110–118 (in Russian).

[Agmon S. (1954)] The relaxation method for linear inequalities. *Canad. J. Math.*, 6 (3), 382–392.

[Arrow K.J., Hurwicz L., and Udzava H. (1958)] *Studies in Linear and Nonlinear Programming*. Stanford University Press, Stanford, California.

[Bakushinsky A. and Goncharsky A. (1994)] *Ill-Posed Problems: Theory and Applications*. Kluwer Acad. Publ., Dordrecht etc.

[Berdyshev V.I. and Petrak L.V. (1999)] *Approximation of Functions, Compression of Numerical Information, Applications*. Inst. Mat. Mekh., Ural. Otdeleniye RAS, Ekaterinburg (in Russian).

[Bourdet J.A., Ayoub J.A., and Pirard Y.M. (1989)] Use of pressure derivative in well test interpretation. *SPE Formation Evaluation*, June Paper SPE 12777, 293–302.

[Bregman L.M. (1965)] Finding the general point of convex sets by the method of successive projecting. *Dokl. AN SSSR*, 162 (3), 487–490 (in Russian).

[Browder F. (1967)] Convergence of approximation to fixed point of nonexpansive nonlinear maps in Banach spaces. *Arch. Ration. Mech. Anal.*, 24 (1), 82–90.

[Clark F. (1988)] *Optimization and Nonsmooth Analysis*. John Wiley and Sons, New York etc.

[Coats D.L., Rapoport L.A., McCord J.R., and Drews W.P. (1964)] Determination of aquifer influence functions from field data. *JPT*, 1417, December, Trans, AIME, 231.

[Dem'yanov V.F. and Vasil'ev L.V. (1981)] *Nondifferentiable Optimization.* Nauka, Moscow (in Russian).

[Dotson W.G. (1970)] On the Mann iterative process. *Trans. Amer. Math. Soc.*, 149 (1). 65–73.

[Eberlein W.F. (1949)] Abstract ergodic theorems and weak almost periodic function. *Trans. Amer. Math. Soc.*, 67, 217–240.

[Edwards R.E. (1962)] *Functional Analysis. Theory and Applications.* Holt Rinehart and Winston, New York etc.

[Engl H.W., Hanke M., and Neubauer A. (1996)] *Regularization of Inverse Problems.* Kluwer Acad. Publ., Dordrecht etc.

[Eremin I.I. (1962)] Iterative method of Chebyshev approximations of improper systems of linear inequalities. *Dokl. AN SSSR*, 143 (6), 1254–1256 (in Russian).

[Eremin I.I. (1965a)] Generalization of the Motzkin–Agmon relaxation method. *Uspekhi Mat. Nauk*, 20 (2), 183–187 (in Russian).

[Eremin I.I. (1965b)] Relaxation method for solving an inequality system with convex functions in the left-hand sides. *Dokl. AN SSSR*, 160 (5), 994–996 (in Russian).

[Eremin I.I. (1966a)] On the penalty method in the convex programming. *Abstracts of International Matem. Congress*, Section 14, Moscow.

[Eremin I.I. (1966b)] On some iterative methods in the convex programming. *Ekonomika i Matem. Metody*, 2 (6), 870–886 (in Russian).

[Eremin I.I. (1966c)] On inequality systems with convex functions in the left-hand sides. *Izv. AN SSSR*, seria. Matem., 30 (2), 265–278 (in Russian).

[Eremin I.I. (1967)] On the penalty method in the convex programming. *Kibernetika*, 4, 63–67 (in Russian).

[Eremin I.I. (1968a)] Methods of Fejér approximations in the convex programming. *Matem. Zametki*, 3 (2), 217–234 (in Russian).

[Eremin I.I. (1968b)] On the rate of convergence in the method of Fejér approximations. *Matem. Ekonomika*, 4 (1), 53–60 (in Russian).

[Eremin I.I. (1969a)] On the general theory of Fejér mappings. *Matem. Zapiski Ural'skogo Gos. Universit.*, 7 (2), 50–58 (in Russian).

[Eremin I.I. (1969b)] Application of the method of Fejér approximations to solving problems of the convex programming with nonsmooth constraints. *Zhurn. Vych. Mat. Mat. Fiz.*, 9 (5), 1153–1160 (in Russian).

[Eremin I.I. (1999)] *Theory of Linear Optimization*. Izd. "Ekaterinburg", Ekaterinburg (in Russian).

[Eremin I.I. and Astaf'ev N.N. (1976)] *Introduction into the Theory of Linear and Nonlinear Programming*. Nauka, Moscow (in Russian).

[Eremin I.I. and Mazurov V.D. (1979)] *Nonstationary Processes of Mathematical Programming*. Nauka, Moscow (in Russian).

[Eremin I.I. and Popov L.D. (2002)] Parallel Fejér methods for strongly structured systems of linear inequalities and equations. *Algoritmy i programmniye sredstva parallel'nykh vyxhisleniyi*, Inst, Mat. Mekh., Ural. Otdeleniye RAS, 6, 149–175 (in Russian).

[Eremin I.I. and Sokolinskaya I.M. (2002)] Fejér iterative processes for ill-posed problems of linear programming. *Matemat. strustury i modelirovaniye*, Izd. OmGU, Omsk, 9, 10–26 (in Russian).

[Fejér L. (1922)] Über die Lage der Nullstellen von Polynomen, die aus Minimumforderungen gewisser Art entspringen. Math. Ann., 85 (1), 41–48.

[Genel A. and Lindenstrauss L. (1975)] An example concerning fixed point. *Isr. J. Math.*, 22 (1), 81–86.

[Gilyazov S.F. and Gol'dman N.L. (2000)] *Regularization of Ill-Posed Problems by Iterative Methods*. Kluwer Acad. Publ., Dordrecht etc.

[Groetsch C.W. (1993)] *Inverse Problems in the Mathematical Sciences*. Vieweg, Braunschweig.

[Gurin L.G., Polyak B.T., and Raik E.V. (1967)] Projection methods for finding general point of convex sets. *Zhurnal Vych. Mat. Mat. Fyz.*, 7 (6), 1211–1228 (in Russian).

[Halperin B. (1967)] Fixed points of nonexpansive maps. *Bull. Amer. Math. Soc.*, 73 (6), 957–961.

[Hanke M., Neubauer A., and Scherzer O. (1995)] A convergence analysis of the Landweber iteration for nonlinear ill-posed problems. *Numer. Math.*, 72, 21–37.

[Hoffman A.J. (1952)] On approximate solutions of systems of linear inequalities. *J. Res. Nat. Bureau Stand.*, 49 (4), 263–265.

[Ivanov V.K., Vasin V.V., and Tanana V.P. (2002)] *Theory of Linear Ill-Posed Problems and Its Applications*. USP, Utrecht, The Netherlands.

[Ioffe A.D. and Tikhomirov V.M. (1974)] *Theory of Extremal Problems*. Nauka, Moscow (in Russian).

[Kaczmarz S. (1937)] Angenäherte Auflösung von Systemen linearer Gleichungen. *Bull. Inst. Acad. Pol. Sci. Leet.*, 35 (4), 355–357.

[Kolmogorov A.N. and Fomin S.V. (1976)] *Elements of the Theory of Functions and Functional Analysis*. Nauka, Moscow (in Russian).

[Krasnosel'sky M.A., Zabreyiko G.M., Zabreyiko P.P., et. al. (1969)] *Approximate Solving the Operator Equations*. Nauka, Moscow (in Russian).

[Lavrent'ev M.M. (1962)] *On Some Ill-Posed Problems of Mathematical Physics*. Izd. Sib. Otdelen. AN SSSR, Novosibirsk (in Russian).

[Leonov A.S. (1999)] Piece-wise uniform regularization of two-dimensional ill-posed problems with discontinuous solutions: numerical analysis. *Zhurnal Vych. Mat. Mat. Fiz.*, 39 (12), 1934–1944 (in Russian).

[Martinet B. (1972)] Determination approachee d'un point fixe d'une application pseudo-contractante. Cas de l'application prox. *C. R. Acad. Sci. Paris.*, Seria A-B, 274, A163–A165.

[Maruster S. (1977)] Quasi-nonexpansivity and two classical methods for solving nonlinear equations. *Proc. Amer. Math. Soc.*, 62 (1), 119–123.

[Merzlyakov Yu.I. (1962)] On one relaxation method for solving systems of linear inequalities. *Zhurnal Vych. Mat. Mat. Fiz.*, 2 (3), 482–487 (in Russian).

[Moreau J.J. (1965)] Proximite et dualite dans un espace Hilbertien. *Bull. Soc. Math. France.*, 93 (2), 273–299.

[Motzkin T.S. and Schoenberg J.J. (1954)] The relaxation method for linear inequalities. *Canad. J. Math.*, 6 (3), 393–404.

[Neubauer A. and Scherzer O. (1995)] A convergence rate result for a steepest descent method and minimal error method for the solution of nonlinear ill-posed problems. *J. Anal. Appl.*, 14 (2), 369–377.

[Polyak B.T. (1969)] Minimization of nonsmooth functionals. *Zhurnal Vych. Mat. Mat. Fiz.*, 9 (3), 509–521 (in Russian).

[Polyak B.T. (1983)] *Introduction into Optimization*. Nauka, Moscow (in Russian).

[Robinson J. (1951)] An iterative method of solving a game. *Annals of Mathematics*, 54 (2), 296–301.

[Rockafellar R.T. (1970)] *Convex Analysis*. Princeton University Press, Princeton, New Jersey.

[Scherzer O. (1995)] Convergence criteria of iterative methods based on Landveber iteration for solving nonlinear problems. *J. Math. Anal. and Appl.*, 194, 911–933.

[Schroter T., Hollander F., and Gringarten C. (2001)] Deconvolution of well test data analysis as a nonlinear total least squares problem. *2001 SPE Annual Technical Conference and Exibition*, New Orleans, 30 Sep – 3 Oct, Paper 71574.

[Shor N.Z. (1979)] *Methods for Minimization of Nondifferentiable Functions and Their Applications*. Naukova Dumka, Kiev (in Russian).

[Tikhonov A.N., Leonov A.S., and Yagola A.G. (1995)] *Nonlinear Ill-Posed Problems*. Nauka, Moscow (in Russian).

[Vainberg M.M. (1972)] *Variational Method and Method of Monotone Operators*. Nauka, Moscow (in Russian).

[Vasin V.V. (1981)] General scheme of discretization of regularizing algorithm in Banach spaces. *Dokl. AN SSSR*, 256 (2), 271–275 (in Russian).

[Vasin V.V. (1982)] Discrete approximation and stability in extremal problems. *Zhurnal Vych. Mat. Mat. Fyz.*, 22 (4), 269–280 (in Russian).

[Vasin V.V. (1988)] Iterative methods for solving ill-posed problems with *a priori* information in Hilbert spaces. *Zhurnal Vych. Mat. Mat. Fyz.*, 28 (7), 971–980 (in Russian).

[Vasin V.V. (1995)] Iterative approximation of solution in a finite problem of moments. *Dokl. AN SSSR*, 318 (5), 1042–1045 (in Russian).

[Vasin V.V. (1998)] On convergence of methods of the gradient type for nonlinear equations. *Doklady RAS*, 359 (1), 7–9 (in Russian).

[Vasin V.V. (2002)] Regularization and iterative approximation for linear ill-posed problems in the space of functions of bounded variation. *Proc. Stekl. Inst. Math.*, Suppl. 1, S225–S229.

[Vasin V.V. and Ageev L.A. (1995)] *Ill-Posed Problems with A Priori Information*. USP, Utrecht, The Netherlands.

[Vasin V.V., Gribanov K.G., Zhakharov V.I. et al. (2006)] Regular methods of solution of ill-posed problems for remote sensing of Earth atmosphere using high-resolution spectrometry. Proc. SPE, 6880, 65800T (Dec. 12, 2006).

[Vasin V.V. and Mokrushin A.A. (2000)] Iterative processes of the Gauss–Newton type for ill-posed operator equations. *Doklady RAS*, 371 (1), 35–37 (in Russian).

[Vasin V.V. and Serezhnikova T.I. (2001)] Two-stage method of approximation of nonsmooth solutions and reconstruction of noised images. *Avtomatika i telemekhanika*, 2, 126–135 (in Russian).

[Vogel C.R. (2002)] *Computational Methods for Inverse Problems*. SIAM, Philadelphia.

[Vogel C.R. and Hanke M. (2002)] Two-level preconditioners for regularized inverse problems II: Implementation and numerical results. Preprint, 1998. Internet, `ftp.math.montana.edu/pub/vogel`.

Notations

$\mathbb{R}^n$	real Euclidean space
$\mathcal{H}$	real Hilbert space
$\mathcal{X}$	real Banach space
$(\cdot, \cdot)$	inner product in $\mathbb{R}^n$ or $\mathcal{H}$
$\\|\cdot\\|$	Euclidean norm
I	identity operator or identity matrix
$A^\top$	matrix transposed to A
A^*	operator conjugated to A
$\mathcal{D}(T)$	domain of definition of operator T
$R(T)$	domain of values of operator T
$\\|A\\|$	norm of operator (matrix) A
$\sigma(A)$	spectrum of operator A
$2^{\mathcal{X}}$	set of all subsets of set $\mathcal{X}$
$\mathrm{Fix}(T)$	set of fixed points of operator T
$\bar{S}_r(\theta)$	closed ball with radius r and center at point θ
$[x_0, x_1]$	segment: $[x_0, x_1] =$ $\{x : x_\lambda = \lambda x_0 + (1-\lambda)x_1,\ 0 \le \lambda \le 1\}$
$\mathrm{conv}(S)$	convex hull of set S
$\mathrm{int}\, Q$ or Q^0	interior of set Q
$\mathrm{dom}\, f$	effective domain of functional f: $\mathrm{dom}\, f : \{x \in \mathcal{H} : f(x) < \infty\}$
$f^+(x)$	positive part of functional $f(x)$: $f^+(x) = \max\{f(x), 0\}$
$\partial f(x)$	subdifferential of convex functional
$\nabla f(x)$	gradient of functional

$\mathrm{Lin}\{x_1, x_2, \dots, x_k\}$	linear hull cretated by elements $\{x_1, x_2, \dots, x_k\}$
$P_Q(x)$	metric projection of x onto set Q
$(P_Q)_k$	metric projection onto set Q in space with variable norm
$P_Q^\lambda(x)$	metric projection onto set Q with relaxation coefficient λ
$\rho_M(x)$ or $\lvert x - M\rvert$	distance from point x to set M
S_Q^f	*prox*-mapping: $S_Q^f : v \to \arg\min\{f(x) + \frac{1}{2}\,\|x - v\|^2 :\ x \in Q\}$
$\arg\min_{x\in Q} f(x)$	set of point of global minimum $f(x)$ on set Q
$\arg\min_{x\in Q} f(x)$	point of global minimum $f(x)$ on set Q
arg (6.2)	optimal set of problem (6.2)
opt (6.2)	optimal value of problem (6.2)
L	problem of linear programming (LP) (Section 4, Chapter IV)
L^*	problem dual to problem LP (Section 4, Chapter IV)
0	zero or zero vector (element)
$\mathcal{K}_M$	class of weakly M-Fejér (M-quasi-nonexpansive) mappings (Section 1, Chapter I)
$\mathcal{F}_M$	class of M-Fejér (strictly M-nonexpansive) mappings (Section 1, Chapter I)
$\mathcal{P}_M^\nu$	class of M-pseudo-contractive (or strongly M-Fejér) mappings (Section 1, Chapter I)
$\mathcal{K}$	class of nonexpansive (nonexpanding) mappings (Section 1, Chapter I)
$\mathcal{P}^\nu$	class of pseudo-contractive mappings (Section 1, Chapter I)
"$\to$"	sign of strong convergence of sequence
"$\rightharpoondown$"	sign of weak convergence of sequence
"$\sim$"	sign of identity

Index